IMS

A New Model for
Blending Applications

OTHER TELECOMMUNICATIONS BOOKS FROM AUERBACH

Active and Programmable Networks for Adaptive Architectures and Services
Syed Asad Hussain
ISBN: 0-8493-8214-9

Ad Hoc Mobile Wireless Networks: Principles, Protocols and Applications
Subir Kumar Sarkar, T.G. Basavaraju, and C. Puttamadappa
ISBN: 1-4200-6221-2

Comprehensive Glossary of Telecom Abbreviations and Acronyms
Ali Akbar Arabi
ISBN: 1-4200-5866-5

Contemporary Coding Techniques and Applications for Mobile Communications
Onur Osman and Osman Nuri Ucan
ISBN: 1-4200-5461-9

Context-Aware Pervasive Systems: Architectures for a New Breed of Applications
Seng Loke
ISBN: 0-8493-7255-0

Data-driven Block Ciphers for Fast Telecommunication Systems
Nikolai Moldovyan and Alexander A. Moldovyan
ISBN: 1-4200-5411-2

Distributed Antenna Systems: Open Architecture for Future Wireless Communications
Honglin Hu, Yan Zhang, and Jijun Luo
ISBN: 1-4200-4288-2

Encyclopedia of Wireless and Mobile Communications
Borko Furht
ISBN: 1-4200-4326-9

Handbook of Mobile Broadcasting: DVB-H, DMB, ISDB-T, AND MEDIAFLO
Borko Furht and Syed A. Ahson
ISBN: 1-4200-5386-8

The Handbook of Mobile Middleware
Paolo Bellavista and Antonio Corradi
ISBN: 0-8493-3833-6

The Internet of Things: From RFID to the Next-Generation Pervasive Networked Systems
Lu Yan, Yan Zhang, Laurence T. Yang, and Huansheng Ning
ISBN: 1-4200-5281-0

Introduction to Mobile Communications: Technology, Services, Markets
Tony Wakefield, Dave McNally, David Bowler, and Alan Mayne
ISBN: 1-4200-4653-5

Millimeter Wave Technology in Wireless PAN, LAN, and MAN
Shao-Qiu Xiao, Ming-Tuo Zhou, and Yan Zhang
ISBN: 0-8493-8227-0

Mobile WiMAX: Toward Broadband Wireless Metropolitan Area Networks
Yan Zhang and Hsiao-Hwa Chen
ISBN: 0-8493-2624-9

Optical Wireless Communications: IR for Wireless Connectivity
Roberto Ramirez-Iniguez, Sevia M. Idrus, and Ziran Sun
ISBN: 0-8493-7209-7

Performance Optimization of Digital Communications Systems
Vladimir Mitlin
ISBN: 0-8493-6896-0

Physical Principles of Wireless Communications
Victor L. Granatstein
ISBN: 0-8493-3259-1

Principles of Mobile Computing and Communications
Mazliza Othman
ISBN: 1-4200-6158-5

Resource, Mobility, and Security Management in Wireless Networks and Mobile Communications
Yan Zhang, Honglin Hu, and Masayuki Fujise
ISBN: 0-8493-8036-7

Security in Wireless Mesh Networks
Yan Zhang, Jun Zheng, and Honglin Hu
ISBN: 0-8493-8250-5

Wireless Ad Hoc Networking: Personal-Area, Local-Area, and the Sensory-Area Networks
Shih-Lin Wu and Yu-Chee Tseng
ISBN: 0-8493-9254-3

Wireless Mesh Networking: Architectures, Protocols and Standards
Yan Zhang, Jijun Luo, and Honglin Hu
ISBN: 0-8493-7399-9

AUERBACH PUBLICATIONS
www.auerbach-publications.com
To Order Call: 1-800-272-7737 • Fax: 1-800-374-3401
E-mail: orders@crcpress.com

IMS

A New Model for Blending Applications

Mark Wuthnow ■ Matthew Stafford ■ Jerry Shih

CRC Press
Taylor & Francis Group
Boca Raton   London   New York

CRC Press is an imprint of the
Taylor & Francis Group, an **informa** business
AN AUERBACH BOOK

Nokia and Nokia 5800 XpressMusic are trademarks or registered trademarks of Nokia Corporation.

Auerbach Publications
Taylor & Francis Group
6000 Broken Sound Parkway NW, Suite 300
Boca Raton, FL 33487-2742

© 2010 by Taylor and Francis Group, LLC
Auerbach Publications is an imprint of Taylor & Francis Group, an Informa business

No claim to original U.S. Government works

Printed in the United States of America on acid-free paper
10 9 8 7 6 5 4 3 2 1

International Standard Book Number: 978-1-4200-9285-1 (Hardback)

Library of Congress Cataloging-in-Publication Data

Wuthnow, Mark.
    IMS : a new model for blending applications / / Mark Wuthnow, Jerry Shih, and Matthew Stafford.
        p. cm. -- (Informa telecoms & media ; 9)
    Includes bibliographical references and index.
    ISBN-13: 978-1-4200-9285-1 (alk. paper)
    ISBN-10: 1-4200-9285-5 (alk. paper)
    1. Internet telephony. 2. Cellular telephone systems. 3. Convergence (Telecommunication) I. Shih, Jerry. II. Stafford, Matthew. III. Title.

TK5105.8865.W87 2009
384.5'3--dc22   2009004373

Visit the Taylor & Francis Web site at
http://www.taylorandfrancis.com

and the Auerbach Web site at
http://www.auerbach-publications.com

Dedication

To my beloved wife Debbie whose constant support enabled me to complete
this work and to my three joys—Stephanie, Daniel, and Nathan.

MW

Contents

SECTION II IMS SIGNALING PRIMER

SECTION III IMS/OMA-BASED ENABLERS

Foreword

The world of telecommunications is undergoing a rapid and dramatic change in the fundamentals of how applications are designed, developed, and delivered. For many years, new telecom applications were developed using closed and specialized architectures that required multiyear efforts and resulted in stovepipes that could not be leveraged further. Today, several factors are conspiring to change this. The success of the Internet has made IP-based packet networks the de facto basis for future communications services. The long-running promise of converged voice, video, and data is finally becoming reality. An increasingly mobile, "always on" society has ratcheted up the adoption rate of new applications to a pace that was unimaginable in the days of the old Public Switched Telephone Network (PSTN). Further, an increasingly small world means that users will no longer be served by a single provider in a single location.

The IP Multimedia Subsystem (IMS) is reaching maturity as a standard and beginning to see widespread deployment amid all of these forces. Although it is not a silver bullet solution to all of the current challenges, IMS is likely to play a major role in the coming years. Given this fact, it is important that anybody working in and around these converging fields should have a general understanding of the role that IMS will play and how it will change the way new applications are designed and deployed in telecommunications networks.

For the last several years, we have been using IMS as a vehicle for bringing the telecommunications world back into the curriculum for students at Georgia Institute of Technology. Through new courses and a student research competition in close cooperation with AT&T, we have challenged students in both engineering and business disciplines to bring their best ideas for new mobile applications to fruition using the IMS architecture. Most students have naturally viewed the Internet and the Web as the obvious place to launch new applications. The traditional telecommunications networks present barriers to innovation that keep the youngest and brightest from considering them as a place to launch new applications. With IMS, there is an opportunity to bring the same type of rapidly launched, blended applications to a marketplace that was previously closed.

This book is written to present IMS in a way that is both approachable and meaningful to students and professionals with both technical and nontechnical backgrounds. The authors have provided readers with the information they need to understand the IMS architecture and how it will change the way applications are delivered to consumers in the new telecommunications world. With a collective 50-plus years of experience in the field and first-hand knowledge of some of the earliest commercial deployments of IMS, the authors have a broad understanding of the opportunities that IMS presents as well as the challenges to its success.

Russell Clark, PhD
Ron Hutchins, PhD
Georgia Tech Research Network Operations Center (GT-RNOC)
Georgia Institute of Technology

Foreword

The IP Multimedia Subsystem was initiated by carriers facing the need to expand their service offerings into the multimedia realm. Carriers and equipment vendors created the IP Multimedia Subsystem (IMS) to advance their abilities to provide new, innovative, multimedia services while still retaining the customer service and quality that telephony carriers have been providing their customers. The industry faced the need to couple IP technology with the ability to provide the end user with managed features and network quality. *IMS: A New Model for Blending Applications* describes this fusion of IP and telephony, and shows how the values and common assets of the telephony world are applied to the flexibility of the Internet world to create a multimedia user experience not previously found in either. Indeed, the book describes the evolution of basic telephony into IP-based multimedia sessions where voice is a component of a multimedia session that can be blended with applications from the Internet.

This work first takes the reader through the mechanics of the IMS to build an understanding of the building blocks of the system. As they describe the IMS, the authors remind us that IMS is not a protocol; instead, the IMS embodies an environment of critical functions, which include control, subscriber data management, access management, bearer handling, charging, etc. The authors show us that the elements of the IMS can be combined in multiple ways to build multimedia applications. Some of these applications are extensions of telephony; many are creations well beyond the sphere of traditional telephony. They also show that one may apply various parts of the IMS to create an application.

The power of the IMS lies in the ability to apply common elements to diverse problems and products. This work shows us many real-life examples illustrating the capability of the IMS to facilitate rapid creation of new and powerful applications. Another fundamental tenet of IMS is its access independence plus the ability to support interdomain interconnection. The example applications illustrate use cases that show how the convergence of these different access networks can lead to a richer customer service experience. IMS natively enables a common environment for a user across multiple access technologies including wireless and wired access.

Those who wish to understand how and why IMS will succeed will want to examine the issue from both the user and carrier perspectives. The book closes with sets of use cases that illuminate the motivation for IMS. Use cases are shown for the consumer and business markets. The authors are highly qualified in this endeavor as they have extensive experience in the industry from the equipment provider, carrier, and standards perspectives.

Mark Wuthnow, Matt Stafford, and Jerry Shih are well respected within the telecommunications industry and have been at the forefront of IMS for the operator industry for several years. Collectively, they cover many technologies related to IMS or interacting with IMS. Their expertise will provide readers with a unique insight and perspective on IMS.

Paul Mankiewich
CTO
Alcatel-Lucent America

Preface

For over a hundred years of modern telecommunication history, public telephone communication has taken place over some form of a circuit switch transmission, which has evolved in its own right from analog to digital and from fixed to mobile networks. Now with the coming of IP Multimedia Subsystem (IMS) technology, which merges the Internet world with the telecommunication world, the days of circuit switch technology as a principal mechanism for telecommunication systems are numbered. IMS has been a major work effort of the 3GPP standards bodies for several years now. With the ratification of the first IMS standards in the mobile network environment (in 3GPP Release 5.0, circa 2002), IMS has subsequently been endorsed by other standards bodies, most notably, 3GPP2 representing the other dominant mobile technologies, as well as ETSI TISPAN representing fixed carrier networks. Today, arguably, IMS has reached the critical mass of being a universal set of network services standards enabled by IP transport and SIP signaling that can span across mobile and fixed network environments in an ubiquitous way. Justifiably, IMS represents a quantum leap from the hundred-year-old circuit-switched network and individual services paradigm toward a packet-switched network fabric based on IP and SIP mechanisms enabling services and applications that are capable of carrying multimedia content with dynamic user control and on-demand reconfiguration.

Many good reference books have been written that go into the intricate details of the IMS standard. To repeat this approach here in this book would add little value to our industry as a whole. For those interested in going down to the "bit and byte" level, we would refer you to one of the many well-written books on this subject. Instead, the focus we would like to take in this book is to provide a better understanding of the application layer and the types of applications you can implement with an IMS network. The goal we would like to achieve will be to go beyond just an explanation of the standard itself and instead address how IMS-based services could actually be deployed in an operator's network, taking into account some real-world practicalities.

For those unfamiliar with IMS, we start the first part of this book with an overview of what IMS is and the components that make up IMS. New readers will understand how the IMS network came about, the different parts of IMS, and a basic introduction to the signaling and call flows to establish an IMS session. A higher level approach is taken here for those not needing the details nor having the time to delve into the details; however, sufficient details will be provided to provide a solid foundation to understand the infrastructure upon which IMS applications will be deployed.

Part II provides a brief tutorial on several of the key signaling standards that have been defined for implementation of an end-to-end service making use of IMS. While the well-known SIP protocol (along with SDP) is discussed, other signaling standards that are necessary to implement an IMS service are also reviewed.

Part III deals with enabling technologies (or enablers) that would be used in an IMS to complement an application. Much of the work done in this area is conducted under the auspices of the Open Mobile Alliance (OMA), which has agreed with 3GPP to address the application plane of the IMS while 3GPP focuses more on the infrastructure (control and bearer) plane. OMA has addressed numerous enablers, of which we will attempt to explain some of the key ones as they relate to blending applications, discussing what they are and how they are implemented.

Finally, Part IV will look at how one can take the enablers defined by OMA, place them in an IMS network defined by 3GPP, add creativity from the developer community, and end up offering new services to the consumer or enterprise customer. One key aspect that will be explored in this part is a key foundational advantages of IMS, that is, the ability to combine multiple enablers and multiple applications into a single end-user service. These concepts are conveyed to the reader through a series of real-life vignettes that describe actual scenarios of how end users could actually use different IMS applications in the course of their day. This approach was taken to mimic the way an operator's marketing organization might go about building the justification in a business case to deploy an IMS and its related enablers and applications. For each of these scenarios, the components of the example IMS service are then broken down and an explanation is laid out showing how the service could be implemented.

This book should be of value to a broad range of engineering, marketing, sales, and managerial professionals as well as to the introductory student who needs a basic understanding of what IMS is along with an understanding of the richness and robustness of the application suite that it offers.

Mark Wuthnow
mark.wuthnow@att.com

Jerry Shih
jerry.shih@att.com

Matthew Stafford
mathew.stafford@att.com

Acknowledgments

To develop a new communications paradigm along with its associated technology, such as the IP Multimedia Subsystem (IMS), requires the cooperation of many organizations representing the operator, vendor, and academic communities in order to make it successful in the marketplace. In a similar manner, this book would not have come together without the support of many fine individuals from these communities. The authors acknowledge and thank each of these individuals for their contributions by providing insightful comments, suggestions, and assistance in the writing of this book.

Alcatel-Lucent:
  Bill Bushnell
  Jim Calme
  Bob Kamp
  Rick Lewis
  Greg Thompson

AT&T:
  Gurmeet Bhatia
  Jason Brown
  Virgilio Corral
  Jasminka Dizderavic
  Steve Frew
  Sreenivasa Gorti
  Nick Huslak
  Megan Klenzak
  Kennie Kwong
  Ileana Leuca
  Jeff Mikan
  Rich Schmidt
  Tye Schriever

Scott Swanburg
Kelly Williams

Ericsson:
  Kjell Johansson
  Lee O'Neal

Georgia Tech:
  Russ Clark

jNetX:
  Steve Lasko
  Matthieu Loreille
  Andre Moskal

Nokia:
  Joel Quejada

NSN:
  Luc Cools
  Mario Muth
  Sami Pekkala
  Cary Wilson

Sony-Ericsson Mobile:
  Mark Kokes
  Jonathon Lohr
  Michael Maddux
  Martin Trively

Tekelec:
  Stan McConnell
  Adam Roach
  Robert Sparks

Special acknowledgment and thanks go to Bill Rosenberg (AT&T) who helped out a bunch of network guys and brought his many years of experience and insight in working with mobile devices to the writing of Chapter 9 for us.

The Authors

Mark Wuthnow, currently part of AT&T's architecture and planning organization, brings over 20 years of telecommunications experience to this work. He started his career at AT&T Bell Laboratories (now Alcatel-Lucent Bell Laboratories) where he led the systems engineering teams responsible for the first three development releases of the ISDN Primary Rate Interface (PRI) for the 5ESS switch. He later joined Southwestern Bell Technology Resources Inc. (now AT&T Laboratories) where he has been working in the wireless arena for over 15 years. Mark has held various positions and has contributed to the various wireless subsidiaries of AT&T throughout their multiple mergers (including Southwestern Bell Mobile Systems, Cingular Wireless, and AT&T Mobility). From his early projects of bringing SS7 and digital cellular technologies into the mobile network to heading the IMS evaluation team that led to Cingular Wireless' decision to deploy IMS, Mark has enjoyed working in the forefront of bringing new technologies into the wireless world. He is currently exploring how the various OMA enablers can be realized across AT&T's entire network. In addition, he was previously involved in the T1 standards bodies developing wireless messaging standards. Mark received both his bachelor and master of science degrees in electrical engineering from the University of Texas at El Paso where he graduated with honors. In addition, he is a senior member of IEEE and holds 12 patents.

Jerry Shih, currently part of AT&T's architecture and planning organization, is representing AT&T in messaging-related industry forums and the activities of standards development organizations. He has been in chair and vice-chair positions in international SDOs in the past, mostly related to messaging service development. He has been involved in OMA service enablers' development since 2003 and is currently active in OMA-converged IP messaging and converged address book service enablers' development.

Jerry Shih started his telecom career at AT&T Bell Laboratories (now Alcatel-Lucent Bell Laboratories) where he worked on digital PBX call processing software development. In his 20-plus years' telecom career, he has worked for AT&T Bell Laboratories, BellSouth, Southwestern Bell Mobile System, Cingular Wireless, and

AT&T Mobility in different capacities. He has been part of the major telecom evolutions, from ISDN to Intelligent Network to IMS. He holds a master of science degree in computer science from Villanova University.

Matthew Stafford began his telecommunications career in 1996 with SBC Technology Resources Inc. He worked on ATM and Quality of Service in IP networks. Matthew later moved to Cingular Wireless and then to AT&T as a result of its merger with Cingular Wireless. After working on various softswitch projects, Matthew wrote *Signaling and Switching for Packet Telephony* (Artech House, 2004). From 2004 to 2007, he served as vice chair of numbering for GSM North America. The numbering group's main task was to develop ENUM recommendations.

Matthew is currently a member of AT&T's architecture and planning organization. He holds PhD degrees in mathematics (Northwestern University) and operations research and industrial engineering (University of Texas–Austin). Matthew is a member of IEEE.

Bill Rosenberg (Chapter 9 author) is a principal technical architect in AT&T's architecture and planning organization. He has more than 20 years of experience in terrestrial and satellite communications. Bill's career began at TRW Space & Electronics (now Northrop Grumman Space Technology) where he was a systems engineer on satellite communication payloads. He later worked for Arrowsmith Technologies designing wireless communication systems for the company's fleet management product. Bill joined Southwestern Bell Technology Resources (now AT&T Labs) and continued with SBC's wireless division through its various incarnations, including Cingular Wireless and AT&T Mobility.

Bill is supporting AT&T's wireless business by developing new technologies and services. He currently develops new messaging, cell broadcast, and IMS services for GSM and UMTS designs, and contributes to industry standards. Bill played a key role in the development and launch of Video Share, which was AT&T Mobility's first IMS-based service.

Bill received his master of science degree in electrical engineering from the University of Southern California and a bachelor of science degree in electrical engineering from Texas A&M University.

COMPONENTS
OF AN IMS
ARCHITECTURE

I

Chapter 1

Introduction

The rapid proliferation and acceptance of the Internet and at its core, Internet Protocol (IP) technology, has had a profound impact on the way people communicate. Its openness has led to a plethora of innovative (and some not so innovative) new applications. However, with this new model came limitations and poor performance when it came to real-time applications due to the best-effort nature of the Internet and endpoint incompatibility issues. Contrast this with the telecommunications industry, which has evolved over a period of more than one hundred years with a focus on reliability and performance. However, the telecom networks also have a history of being a closed network with new applications coming out at a snail's pace when compared to the Internet model.

So, one might envision merging the best of both of these models together to get a communications network that offers a reliable network that delivers a quality of service (QoS) that will meet the needs of your real-time applications and at the same time provide an environment for new and innovative applications to meet the needs of end users. Also, one could use these services with mobile or fixed telephony devices or any device with broadband access such as a PC or Internet Protocol Television (IPTV) and, correspondingly, over any of their traditional networks. If one did merge IP and telephony, one would come up with the IP Multimedia Subsystem (IMS).

The IMS architecture follows a commonly used three-layer architecture structure. The lower layer, referred to as the *bearer plane* (or sometimes the *transport plane*), is made up of the physical resources necessary for a connection to be established and for the payload (either voice or data) to be carried from the origination point to the termination point. The middle layer is referred to as the *control plane*, which comprises "intelligent" elements that determine whether a user is allowed to

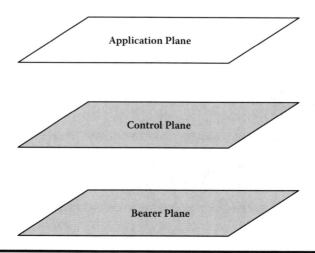

Figure 1.1 IMS layered structure.

use the network and how to route a call or data path (also referred to as *setting up a session*). The top layer is referred to as the *application plane*, which contains all the application's logic and data necessary to offer services to an end user (Figure 1.1).

Although IMS cannot be realistically deployed by a network operator who expects to make a return on their investment without all three layers working together as a unit, it is the application plane with its suite of revenue-generating end-user services that will determine whether deployment of an IMS network is successful or not. It is on this particular plane that this book is focused, examining the new application-layer tools (or enablers) that will be available to operators as well as some specific end-user services. However, with that said, it is critical that to fully understand how to make the best use of the elements in the application plane, there must be an understanding of what the bearer and control planes provide.

Thus, we will examine in Part I the components that make up all three layers (or planes) of the IMS and the pathway taken to arrive at this design. Part II will provide a short primer on the signaling protocols that are defined for use in the application layer. Parts III and IV will focus on the primary theme of this book, the application enablers defined for the IMS and some realistic applications or combinational applications that can be offered to the actual end user. Part III will focus on the working groups within the Open Mobile Alliance (OMA), which is tasked with defining enablers and applications for use in the IMS. Part IV presents a series of use cases and example end-user services for different targeted markets (consumer, enterprise, converged) that could be deployed using IMS. But first, let us take a brief look at what the vision of IMS is trying to convey — namely, the added value that IMS will offer to both service providers and end users, as well as its impact.

1.1 What Is the IMS Vision?

The telecommunications industry is about to undergo a fundamental shift due to IP-based technologies. A sound bite for IMS might be "the convergence of the Internet and the telecom world into a single package." Although this may be an oversimplification, it gets at the crux of the IMS vision: find a way to reduce the cost of the network, and at the same time find a way to offer new and compelling services in a more responsive and economical manner to the marketplace.

However, although the network cost reduction might be justification for service providers to deploy an IMS network, it is really the ability to offer the new and compelling services, which include multimedia-based applications, that will generate the growth revenue stream for operators that will drive IMS deployment. This view coincides with a survey of wireless operators conducted by Canaccord, which showed by a 2-to-1 margin that the driver for IMS is new services (Figure 1.2).

Taking this survey one step further, arguably one can state the key service innovation drivers for the IMS are the following:

- Real-time multimedia interactive services
- Fixed-mobile convergence
- More value in integrating services, enablers, and different access network types than as stand-alone silo applications

Now, by moving to well-known and standardized application programming interfaces (APIs), IMS seeks to bring in an Internet-like application developers model where you have a large pool of developers creating new applications responsive to the marketplace.[1] New applications are expected to come from outside the traditional sources (i.e., network equipment providers [NEPs]). Thus, what IMS provides is an enabling framework to help roll out multiple applications from multiple sources with short market lead times.

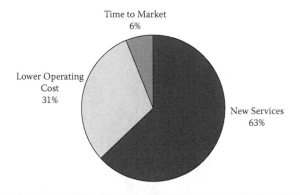

Figure 1.2   Carrier survey—IMS drivers. (Image from Canacord IMS Survey. Used with permission.)

Under the current environment, it is a challenge for operators to deploy more than a couple of new services a year due to the custom work that must be done for each and every new service. With IMS, because of the standardized APIs at all points in the network (including the charging and provisioning interfaces) and on the device, the need for specialized deployments for every new application is greatly reduced. In fact, it is envisioned that new services will roll out on a weekly, if not daily, basis once a deployed IMS network has matured. Under this new model, it is fully expected that the marketplace will see many more specialized service or campaign offerings with possibly just a couple of weeks of actual deployment life.[2] Examples of these limited life services could be campaigns around a special sporting event such as the Super Bowl or World Cup.

Along with a service delivery framework to support the rollout of new applications, an enabler layer is being defined as part of the overall application layer to provide network-specific information in support of a specific application or applications. An enabler is a function that is performed once in the network, and its information is shared among multiple applications for efficiency reasons, lower capital expense (CAPEX), and the ability to share common information among different applications. A simple example would be the enabler Presence. In today's environment, the services Push to Talk (PTT) and Instant Messaging (IM) each have their own presence server (PS). Information about the presence status of a PTT customer is unable to be shared with the IM application. Thus, natural silos are being created between an operator's commercial applications. With a stand-alone presence server as an enabler, an operator needs to deploy but a single presence server as opposed to a presence server per application. Figure 1.3 shows how presence information can be shared across multiple applications, which would allow the example applications

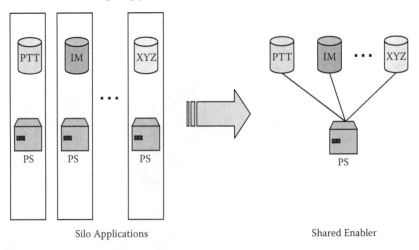

Silo Applications   Shared Enabler

Figure 1.3   Migration to stand-alone enabler.

to have harmonized data so that these applications could be converged into a richer user experience application.

Finally, IMS has been defined as *access network independent* or, to put it a different way, IMS is designed to support any kind of access network; the subscriber is allowed to move across these network boundaries at will and their subscription (and associated subscriber data) follows them. This capability of IMS takes the basic concept of mobility from the wireless world to the next level. It also opens up a whole new definition for converged applications in which applications within a single network can not only be converged (or combined) but can also work across multiple networks as if those networks have been converged together. Figure 1.4 shows a sampling of access networks that a single IMS network can support with a single infrastructure and applications that can be offered across different access networks.

1.2 Added Value of IMS

In Section 1.1, we touched upon the vision laid out for IMS. Now we want to peel back the next layer and look at the added value that IMS provides. If one just takes a superficial glance at IMS, one might conclude that it provides nothing new. With the broadband data capabilities that Third Generation (3G) networks provide, one can access the Internet and get Internet services. End users can surf the Internet, download videos, read e-mail, and do most everything that a "wired" broadband customer can do today. In addition, some of the services being touted as lead IMS services such as Push-to-Talk over Cellular (PoC) and IM are already being offered over 2 or 2.5G networks. So, what is the value addition of IMS over what can be provided today?

In a nutshell, there are four key value propositions that IMS offers over a plain broadband Universal Mobile Telecommunication System (UMTS™) or Wideband CDMA (WCDMA) pipe. These four items are

1. Session control
2. Blending and integration of different applications
3. Quality of service
4. Flexible charging

Let us take a brief look at each of these in turn.

1.2.1 Session Control

Session control allows the IMS network to provide additional features and functionalities to both the originating and terminating ends of a session. A session is the establishment of a path to complete a voice call or a data connection (such as a live

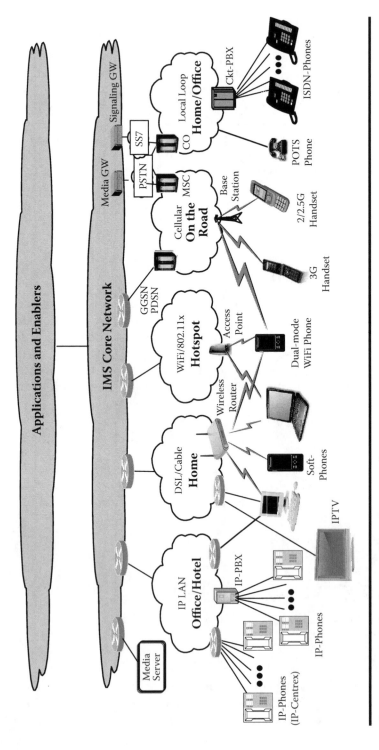

Figure 1.4 Example of IMS multiaccess network support.

video stream). Session control allows one to not only establish a session but to also act on that session during the life of that session (such as to add a new multimedia call leg). The Session Initiation Protocol (SIP) has been selected as the session control protocol for the IMS to establish and to manage multimedia sessions over IP-based networks.

SIP follows a client-server model similar to those used for HTTP (Hypertext Transfer Protocol) and SMTP (Simple Mail Transfer Protocol). In fact, SIP's creators adopted many of its design principles from these two well-known IP protocols. However, SIP does not differentiate between the user-to-network interface (UNI) and the network-to-network Interface (NNI), thus making it an end-to-end protocol. Also, because the client-side device is not a fixed relation with SIP, SIP is able to implement peer-to-peer applications. This peer-to-peer relationship is an important capability as two endpoints could establish applications between themselves and use a service provider's network as a "dumb pipe," given sufficient compatibility between the two end devices and if the two endpoints are fixed (or known) locations.

This is where IMS provides great value in implementing session control. The functions it can provide to the end users to assist the establishment of the session include the following:

- Mobility management to determine the routing information (address) for a mobile endpoint
- Content adaptation for incompatible devices
- Interworking between different access networks
- Application logic not available in device
- Supplemental network information for the application (e.g., Presence status)
- Store and forward capabilities when one user is not connected
- Bridging and multiparty applications
- Quality of service (QoS)

From a network operator's perspective, the ability to provide session control is critical to remaining an integral part of the value chain. Being able to offer to the end user the aforementioned functions allows network operators to provide key capabilities in bringing the best service offerings to their customers. In addition, the ability to provide session control allows network operators to provide necessary security for their network and to obtain information required to properly run their network. Some additional functions that session control offers to the network operator include the following:

- Authentication and authorization of the end user and endpoints
- More efficient allocation of network resources
- Session-charging information
- Level of quality assurance for network-provided applications

1.2.2 Blending and Integration of Different Applications

As mentioned previously, rolling out new applications can be a very time-consuming, complex, and expensive experience for network operators. Even if the different parts of the new application are already deployed, an easy mechanism to bring these different parts together does not exist. However, IMS provides two mechanisms with which to blend applications already deployed in the network to create new end-user services. They are as follows:

1. initial Filter Criteria (iFC)
2. Service broker

The iFC is a set of prioritized trigger points assigned to a subscriber's profile that indicate the order in which multiple application servers (ASs) should be invoked, depending on what services a user has subscribed to. This will cause a set of multiple ASs to be invoked in a prioritized order at the same trigger point. For an example of how the iFC would be used, consider a virtual private network (VPN) customer with four-digit dialing capability who wants to keep a log of all outgoing calls (Figure 1.5). In a pre-IMS scenario, the only way for these two services to be combined together would be for them to be developed together as a single package. This in turn limits the service provider's change-out options if one of the services is not performing properly or if one of the services scales disproportionately compared to the other. It also means the service provider is needlessly paying for a more expensive platform if the majority of their customers subscribe to only one but not the other service.

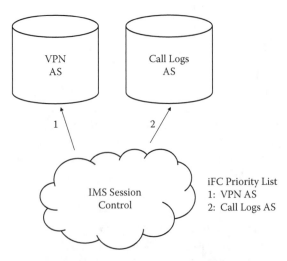

Figure 1.5   iFC example using VPN and call logs.

In our example here, the VPN service is designated with a higher priority than the call log service. The filter criteria are predefined in this order because a number translation is required to obtain the full number prior to logging the call (Figure 1.5).

The service triggering information provided by the iFC is static routing information assigned to a particular user, and is stored as part of his or her subscriber profile data. This type of information is contrasted with the dynamic routing capabilities provided by the service broker. The service broker provides a finer granularity of routing action based on events occurring in the network that are related to the user. On the basis of service rules, the service broker may, for example, change how a user's services are combined, depending on whether the incoming call is a video call and the user is on another call. In this case, the new incoming call may be routed to a special voice mail system that can receive a video voice mail message. We will discuss both the iFC and the service broker in greater detail in Chapter 6.

One final terminological note before we finish this subsection. We are using the term *blend* (or *blending*) to refer to the combination of different applications to provide a new, richer application. Sometimes, this is referred to as *orchestration*. Although this could be a valid term to describe the combination of different applications, "orchestration" also has a foundation in the Web services world and is usually associated with that environment as part of a service-oriented architecture (SOA). To avoid confusion, we chose the term *blend* to avoid placing a limitation on combining applications only in a Web services environment.

1.2.3 Quality of Service

QoS was inherently built into legacy, single-application networks such as the circuit-switched voice network or the cable TV network. In an all-IP network supporting multiple applications, this inherent capability is not present. As most users of the public Internet are aware, it is a best-effort network. For non-real-time applications, this is acceptable. However, for real-time applications, delays in transmission can make a service unacceptable to the end user. On the other hand, giving all sessions (applications) real-time QoS performance (even when not needed) leads to an overbuilding of the service provider's network and drives the cost up for all parties. IMS is designed to provide a QoS with each session that can be assigned based on the subscriber profile or selected application. Third Generation Partnership Project (3GPP™) has defined four QoS classes (also referred to as *traffic classes*):

- Conversational class
- Streaming class
- Interactive class
- Background class

The classes are distinguished from the most delay sensitive (conversational) to the least delay sensitive (background). Thus, conversational class sessions will have the highest priority for network resources, and background class sessions will have the lowest priority for network resources. Conversational and streaming classes are primarily intended for applications requiring real-time traffic. The difference between the two is application sensitivity to delay. Conversational real-time applications such as voice or video telephony are the most delay sensitive and would require the conversational traffic class. The streaming class is intended for a one-way traffic data transport such as an RSS news feed or a YouTube™ video.

The interactive and background classes are intended to be used for traditional Internet-type applications such as e-mail, Web surfing, and Telnet. The distinction between these two classes is that the interactive class would be assigned to interactive-type applications such as interactive e-mail or IM. The background class would be used for downloading files in the background.[3]

1.2.4 Flexible Charging

A flexible charging mechanism is a key way for a network operator to offer a differential value beyond just counting packets. Of course, the operator must be able to support existing charging models, and IMS does support the following existing models:

- Prepaid (online charging)
- Postpaid (offline charging)
- Calling Party Pays
- Called Party Pays
- Data Usage

However, beyond supporting the existing models, IMS charging must be extended to support the key visions for IMS. First, there must be a mechanism to handle the different components of a multimedia application. The following components may be identified in a call detail record (CDR):

- Voice
- Audio (real-time and streaming)
- Video (real-time and streaming)
- Interactive data (e.g., Web surfing)
- Messaging (SMS)
- E-mail
- Unspecified data content stream (network provides only bearer transport)

Each of these multimedia components is identified, and thus can be charged separately.

Next, there must be a way to address additional activities related to a session, such as invoking an enabler (e.g., getting a presence status). Then, one will need to identify the QoS associated with the session. Both of these actions will generate a call detail record (CDR) with appropriate information.

Finally, there must be a mechanism to identify the application server or servers along with the specific application or applications being invoked. These applications can be either within or external to the operator's network. For third-party applications, most operators will place a control gateway between their network and the third party. This control gateway in the IMS architecture is referred to as an *open service access* (OSA) or *Parlay*. In most instances, the charging record for the third-party application would be done here at the OSA Gateway. The OSA Gateway will be further described in Chapter 6.

Of course, all of the proceeding must also work for the network operator's customers roaming into a visited market. Globally unique identifiers are exchanged between network operators and between a network operator and a third-party provider to identify the home and visited networks.

1.3 What Will Be the Impact of IMS?

The telecommunications industry is undergoing a fundamental shift due to IP-based technologies. The three major areas of the industry that will be impacted by this shift are technology, market and services, and the business models. The impact on each of these areas will be discussed separately.

1.3.1 Technology

IP-based technology is fast becoming the basis for all new telecommunications systems. Although in the mobile environment it is still more spectrally efficient to implement a circuit-switched voice call over the air interface, technology advances are expected to overcome this last bastion and pave the way for full end-to-end packet-switched connectivity.[4] Certainly, inside the network, transport is well on its way to becoming fully an IP network. Broadband network operators have already made the end-to-end leap to an all-IP network. As an industry, the telecommunications standards bodies are all converging on IP-based technologies as Internet Engineering Task Force (IETF) standards are being widely adopted in the telecom world. Both the Global System for Mobile Communications (GSM) and Code Division Multiple Access (CDMA) standards bodies are adopting the IETF standards to ensure a common infrastructure network (the CDMA industry segment has given its own name for IMS: Multimedia Domain [MMD]). Vendors have fully adopted IP technology throughout their product portfolio.

The access-independent design of IMS will enable operators to offer convergent applications between fixed and mobile networks. This will enable new types of applications, including the much-vaunted three screen services (IPTV, PC, and mobile device). The simple act of being able to continue a video service session as one moves from coverage on the macrocellular network to a Wi-Fi network and eventually into the home IPTV is a powerful service lure.

1.3.2 Market and Services

With the integration of the Internet into the telecom world (and especially with the mobile wireless world), the customer's expectation is expected to change. Although voice services are still expected to be the mainstay of the ARPU (average revenue per user) collected over the next several years, the percentage of that revenue will see a sizable decrease with a corresponding increase in data-type applications. Data applications will proliferate in many areas, including in combination with voice (also known as a *rich voice call*).

The market should expect to see an adaptation of many traditional Web services to work in a mobile environment, as customers will want those services away from their traditional computer screen. Applications such as mapping directions from Web providers will be combined with the location identification capabilities of a network operator to offer more customer-friendly and mobile-environment-friendly applications. More such examples of combinational applications will be discussed in Part IV of this book.

As mentioned earlier, the advent of the "two-week" service life cycle will greatly codify how the customer views his or her mobile device. Certainly, mass-market, short-lived applications have already appeared in limited form (currently limited because a large effort is still required on the network operator's part to deliver the service). An example here would be the March Madness® (college basketball championship tournament) contest offered by AT&T™. However, these mass-market-type offerings will not bring about a fundamental change in the societal aspects of the marketplace. It will be the community of interest (COI) type of short-lived (or not so short-lived) services that will have more of an impact on the marketplace with its viral implications. COI services are prolific on the Internet and they become enhanced when the mobile component is added to it. Service providers can offer specific targeted services to special-interest groups, such as for people competing in a local marathon or scouting event. Participants can be kept apprised of event schedules; hosts can use the service to contact participants about last-minute changes in schedules or venues. A social networking function can also be created to allow participants to locate other participants for activities such as jogging or sharing a ride to an event.

Of course, all these COI services cannot be maintained by the service provider, because the potential provisioning and updating load could easily overwhelm their staff (think of the Internet analogy). This implies that some sort of a customer self-provisioning mechanism must be activated. As with applications on the Internet, the mobile customer will become more involved in defining and maintaining this next set of viral services.

1.3.3 Business Models

Several new business models appear as network operators and service providers roll out IMS and its supporting infrastructure. Again, certain carryovers from the Internet model will become applicable to the IMS world. New business arrangements and new revenue areas are made available. One well-discussed area is building relationships with the advertising community to open new revenue streams. However, with these new business models also comes new competition that must be accounted for in new business plans.

The access-independent nature of IMS can eliminate the need for separate switching fabric between the wired and the wireless world. This will lead to the convergence of applications across both the fixed and mobile networks. It could also lead to a merging of networks for companies that operate both a fixed and wireless network to reduce their costs. Similarly, it may lead to an interesting partnership arrangement and possibly be a consideration for spurring mergers or takeovers between fixed and wireless operators, and between cable and wireless operators who would like to take advantage of the cost savings associated with a single network infrastructure. At a minimum, one should see more cooperation between the wireless and fixed entities in offering services that cross traditional network boundaries.

The move toward a Service Delivery Platform (SDP) framework enabled by IMS will spur the creation of new applications from both traditional and nontraditional sources. Different industry bodies, including the Tele-Management Forum (TMF), are trying to define how this should be accomplished across multiple technologies. An SDP will open new business and revenue opportunities beyond the obvious goal of just selling new end-user services, such as by enabling new business-to-business (B2B) models in which the operator enables third-party providers to deliver their own services. Many of these sources will be looking for a company to host their new applications as this responsibility is not in their scope of expertise. Here is an opportunity for aggregators or network operators to offer a hosting service for those software companies who do not provide a hosting capability to provide their service. Operators are in a prime position to address this emerging market space as they have the expertise in operating platforms, and they can readily provide billing and collection services for those software companies.

One of the key designs of the SIP protocol used in IMS is its peer-to-peer nature. This means that applications could be implemented in the end-user devices with limited value-added capabilities added by the network operators, except as a big data pipe. Here, the traditional network-centric model that network operators operate under in a pre-IMS environment will run into the device-centric model that comes with SIP-enabled devices. A community of end users (either as a structured community or as a viral community) can run the same application among themselves to get services transparent to the network provider. An example here of a viral community would be Skype™ or Vonage® users placing VoIP calls among themselves. Clever operators will look for creative ways to incorporate both peer-to-peer services as well as network value-added services into their service portfolios. Certain services may obtain a time to market quicker in a peer-to-peer scenario, so the operator is less exposed if the service fails. They can look for ways to provide added value later on as the service becomes commercially feasible. Many different business models will appear in the marketplace. It will be left to the free market to determine what is the best model for a particular market.

Endnotes

1. Although having this large pool of mainly independent developers is a wonderful idea and seductive vendor sales pitch, for operators, their enthusiasm must be tempered with the reality of the need to have a secure network to protect the network and, ultimately, the customer. This reality need will be covered in greater detail in various sections later in this book.
2. This is commonly referred to as the "long tail" model.
3. This is discussed in 3GPP TS 23.107, "Quality of Service (QoS) Concept and Architecture."
4. The cross-over in spectral efficiency from circuit switch to packet switch is expected to occur with the deployment of Long Term Evolution (LTE).

Further Reading

Canaccord Capital Corp. Equity Research, "IMS—The Move to Software," Misek, P., Batrovic, D., and Ofir, E., January 11, 2006.
3GPP TS 23.107, "Quality of Service (QoS) Concept and Architecture."
TMF 519, "Business Agreements," http://www.tmforum.org/TechnicalPrograms/Service DeliveryFramework/4664/Home.html.

Chapter 2

The Path to IMS— Evolution of the Cellular Switching System to 3G

2.1 First Generation

In the beginning (of the cellular world), there was the all-encompassing, self-contained, monolithic switch. It was very big and awkward to modify, and this was not good. The first generation of cellular switches (commonly referred to as mobile switching centers or MSC) went into commercial service in the mid-1980s and kept their basic design for approximately ten years. These early switches had in one single "package" all the capabilities that were needed to provide the necessary functionality to provide cellular switching. These capabilities included (but were not limited to)

- Subscriber profile database (aka, Home Location Register or HLR)
- Incoming and outgoing call functionality
- Call processing logic
- Access and egress ports
- Tones and announcements
- Bridging/transfer
- Application logic
- Feature interaction management
- Call detail records

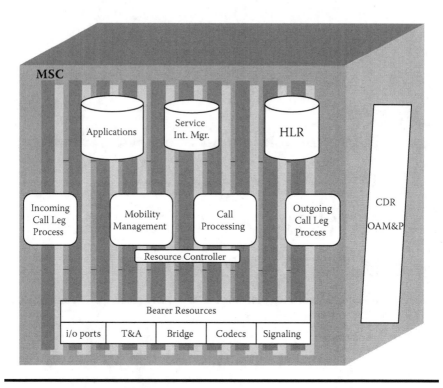

Figure 2.1   First-generation MSC.

Figure 2.1 shows a logical view of these first-generation switches. These switches could be viewed as following a main frame model versus more of a distributed network model with all of call processing and bearer path routing performed on the same platform.

The first-generation switch was the backbone of the analog cellular network. While vendors were able to build products with this model in order to introduce cellular service commercially, it had numerous drawbacks. Its principle drawback was one of scaling. Because of processor or memory constraints, early switches were greatly constrained in the number of subscribers that could be supported per switch. Switches in the late 1980s were size-limited to 50K to 75K subscribers, based on the HLR capacity of the switch. Thus, every time an operator approached these limits a completely new switch was required to be added to their network.[1]

The other principle drawback to these first-generation switches was their complexity in modifying or adding new functionality to them. Because of the tight coupling of all the different capabilities (or subsystems), development of one feature had to be done in lockstep with all the other features to ensure the development of one subsystem did not impact another system. This led to development cycles on the order of three years or more for a new feature to be deployed. And, of course, there were practically no opportunities for an operator to modify any services (except for minimal craft settings availability), much less add their own new service.

2.2 Second Generation

The introduction of SS7 and digital cellular technologies made some early inroads into addressing the two principle drawbacks of the first-generation switches. Both the TIA (covering both TDMA and CDMA) and the Global System for Mobile Communications (GSM) standards bodies defined their architecture to allow for the HLR to be decoupled from the rest of the cellular switch fabric. The TIA specification is described in ANSI-41 (formerly known as IS-41). The GSM specification is described in TS 09.02. The vendor community followed suit with a product commonly referred to as a Stand-Alone Home Location Register (SHLR). Since the SHLR was defined to only process signaling messages, a different approach could be taken in its basic design. Vendors started offering product on more general-purpose computer hardware rather than traditional switching hardware. This allowed for more economy of scale in the basic cost of the platform, plus it took advantage of the fact that the computer industry was more aligned with riding the technology curve of Moore's law than was the telecom switching industry at the time.

Because standardized interfaces were defined between the SHLR and the MSC, both the need to scale and to develop in lock step was eliminated. Each platform could scale independently based on their own unique characteristics (e.g., subscriber data transition rate versus erlangs). Network equipment vendors could develop on their own platform and conduct their testing against a standardized interface without worrying about other adverse impacts to the other platform. This allowed for independent development teams and a quicker time to market for the overall solution.

Figure 2.2 shows a logical view of how the cellular switch is starting to take on aspects of a distributed architecture.[2]

In addition to separating out the HLR from the MSC, initial steps were taken to support new supplementary services external to the MSC while existing services remained with the core MSC. TIA and GSM, respectively, defined the WIN (Wireless Intelligent Network) and CAMEL (Customized Application for Mobile network Enhanced Logic) specifications to support off-board applications. Using a basic call state model, trigger points were defined where call processing was temporarily transferred to an off-board application server (called the *service control point* or *SCP*) for further service logic processing. These temporary transfer points could occur at the origination of call (originating call trigger), during the call (mid-call trigger), or at the end of the call (terminating call trigger). Some examples of CAMEL services that are provided external to the MSC through an SCP would be ring back tones or prepaid service.

Figure 2.3 shows a logical view of the CAMEL architecture which starts to separate out service logic from the MSC into an off-board application server. The gsmSSF refers to the Service Switching Function, which is the functional entity in the MSC that interfaces with the gsmSCF. The gsmSCF is the Service Control Function,[3] which refers to the application server that houses the CAMEL application.

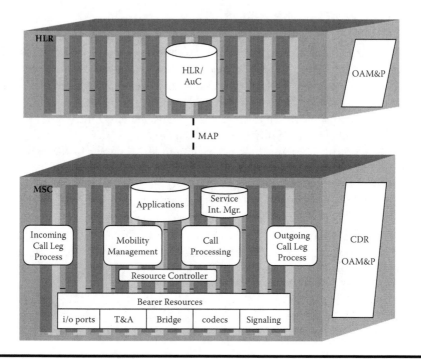

Figure 2.2   Second-generation switching architecture.

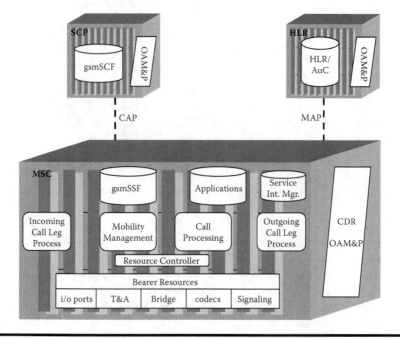

Figure 2.3   CAMEL architecture.

2.3 Third Generation

2.3.1 Release 4—A Transition Architecture to IMS

The third generation of the cellular switching domain was developed in a two-step process. In the Third Generation Partnership Project (3GPP™) Release 4 specifications, the MSC was further divided into the call control functions and the switching (physical resources) function. This type of architecture is commonly referred to as a softswitch architecture. In 3GPP terminology, the call control and the mobility control subsystems are referred to as the MSC server. The switching function is provided by an element called the *Media Gateway* (MGW). The circuit switch core network provided by the Release 4 architecture is intended to be bearer independent and to support both Asynchronous Transfer Mode (ATM) and IP as described in 3GPP TS 23.205, "Bearer-Independent Circuit-Switched Core Network."

Briefly, the MSC server is responsible for the following:

- Handling the mobile's registration for authorization and mobility management
- Providing authentication
- Providing mobile originated and mobile terminated call routing
- Integrated with VLR to hold the subscriber's service data and CAMEL-related data

The MSC server controls the establishment of the bearer channel through the MGW using the Media Gateway Control Protocol (MGCP).

The MGW provides the traditional switching fabric functionality of the MSC. It supports the translation of bearer traffic between and within different types of networks. The MGW is responsible for the following:

- Terminating bearer channel traffic from both circuit-switched (CS) and packet-switched (PS) networks
- Originating bearer channel traffic into both circuit-switched and packet-switched networks
- Providing echo cancellation for circuit-switched connections
- Providing media translation between codecs (e.g., G.711 to AMR)
- Support for conference bridge capabilities
- Optional tones and announcement support

From an end-user point of view, they should notice no difference in having their call connected through either a third-generation Release 4 architecture or through a second-generation architecture. All that the customer will experience is their "business as usual" circuit-switched call. This customer experience in fact was codified in 3GPP TS 23.205, where it states "The users connected to the CS core network shall not be aware whether a MSC server–Media Gateway combination is used or a monolithic MSC is used."

The Release 4 architecture continues the trend of decomposing the monolithic MSC into more distinct components of its basic building blocks. Figure 2.4 shows this continuing trend.

The same benefits that were seen with the SHLR architecture in the previous section can also be applied with the Release 4 architecture. The different components can be independently scaled which means CAPEX dollars can be allocated to specific growth needs versus being diverted to grow capacity in areas not needed. Operators are also starting to have more purchasing flexibility by having these standardized switch components. The telecommunication market has become more competitive as the vendor community does not need the tremendous resources to produce an entire new monolithic switch but, instead, they can chose to focus their resources on smaller individual components. This has greatly improved the price/performance ratio operators pay for their switching products as well as introducing a new set of vendor suppliers of product.

Additional savings are available due to the inherent design of having a distributed architecture. A distributed switch architecture allows for the placement of the MGW (bearer switching) to be placed at the edges of the operator's network to reduce transport charges. Expensive T1 connections are then replaced by much cheaper IP connections as the calls are transported through the network using IP technologies.

Typically, the MGW performs the conversion from a circuit-switched bearer path to a packet-switched bearer path. Thus, the MGW is paving the way for

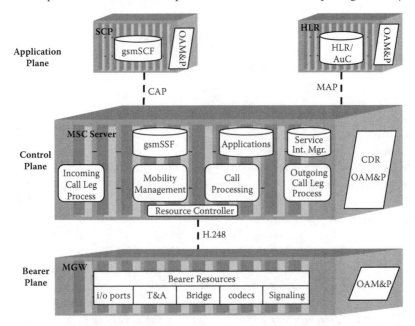

Figure 2.4   3GPP Release 4 high-level architecture.

the beginning of the convergence of the circuit-switched world with the packet-switched world. 3GPP Release 4 is the first network standard defined to work with a Universal Mobile Telecommunication System (UMTS™) network.[4] Starting with Release 4, the 3GPP standards elevate Voice-over-IP (VoIP) as a key capability to be supported. And with VoIP comes the need to provide a quality of service (QoS). ATM was designed to support QoS from its infancy, but IP did not have a similar history. Thus, a number of enhancements have been developed to support QoS over IP such as Resource Reservation Protocol (RSVP), Differentiated Services (DiffServ), and Multi-Protocol Label Switching (MPLS).

2.3.2 Release 5 and Beyond

The IP Multimedia Subsystem (IMS) architecture was first defined with the release of the 3GPP Release 5 specifications and further expanded in the Release 6 and later specifications. In Release 5, the components of our original monolithic switch have been broken down even further into its fundamental building blocks. Thus, what were parts or functions of a Release 4 network element have now been functionally separated into their own network element. A formal three-layer architecture approach has been taken in breaking down the monolithic switch into its fundamental building block components. Figure 2.5 shows our final view of how our original monolithic switch has been realized into a fully distributed architecture. Each

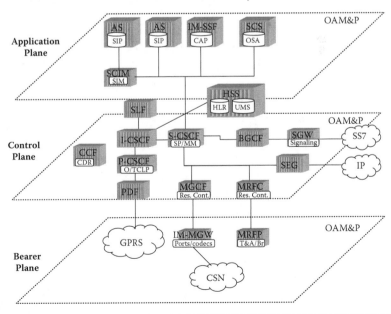

Figure 2.5   IMS component elements and their associate plane.

of our IMS components fall within the three-layer architecture that was defined for the Release 4 Architecture.

Each of these layers and their components will be looked at in more detail in later chapters. However, a brief description follows:

Application plane
- Application Server (AS): Entity that houses a SIP-based application.
- IP Multimedia Service Switching Function (IM-SSF): Interworking point between the IMS and the CAMEL Service Environment.
- Open Service Access Service Capability Server (OSA-SCS): Interworking point between the IMS and the OSA Service Environment.
- Service Capability Interaction Manager (SCIM): An application which performs the role of service interaction manager between different service offerings.

Control plane
- Home Subscriber Server (HSS): Contains all subscriber-related information necessary to support the network to handle calls/sessions for the subscriber.
- Subscriber Locator Function (SLF): Resolution entity to identify the address of the HSS that holds the subscriber data when multiple HSSs are present in a network.
- Call Session Control Function (CSCF): Manages SIP sessions and coordinates with other network elements for session control, feature/service control, and resource allocation. A CSCF may perform one or more of the following roles:
  • Serving CSCF (S-CSCF)—session control point for user equipment (UE) as an originator and terminator
  • Interrogating CSCF (I-CSCF)—the contact point into the UE's home network for other networks
  • Proxy CSCF (P-CSCF)—the contact point into the IMS for the UE
- Breakout Gateway Control Function (BGCF): Selects network to use for PSTN/PLMN interworking.
- Media Gateway Control Function (MGCF): Controls the IMS-MGW.
- Multimedia Resource Control Function (MRCF): Controls the MRCP to establish bearer-related services such as announcements, conference bridging, or bearer transcoding.
- Signaling Gateway (SGW): Interconnects different signaling networks such as IP-based signaling networks with SS7 signaling networks.
- Security Gateway (SEG): Provides security wall between two different security (usually operator) domains.
- Policy Decision Function (PDF): Makes policy decision for a session, authorizes QoS requests.
- Charging Collection Function (CCF): Single reference point from various IMS entities toward the operator's billing system.

Bearer plane
- IMS Media Gateway Function (IMS-MGW): Interworks RTP/IP and PCM bearers (link between the IMS and the CS network) at the request of the MGCF; may also provide tones and announcements.
- Media Resource Function Processor (MRFP): Provides conferencing, transcoding, and announcements resources at the request of the MRFC.

Besides breaking down the Release 4 architecture further into its fundamental components, the IMS architecture furthered two more key concepts from the Release 4 architecture. In Release 4 we saw the introduction of IP in the MGW. With IMS, IP technology becomes the base technology of all components. The Session Initiation Protocol (SIP) becomes the new basis for signaling (replacing SS7 MAP). It should be noted that 3GPP Release 5 was only defined to work with IPv6. However, since many carriers had not upgraded their IP network to support IPv6, most IMS vendors built their first product releases to run over IPv4 (although some did include support for IPv6). In 3GPP Release 6, the IMS standard was expanded to allow for either an IPv4 or an IPv6 network.

In addition, with the Release 6 standard, IMS is defined to work with any access network as opposed to Release 5, which is defined to work only with a GPRS network. Thus, IMS can provide the network infrastructure for various types of networks such as GSM, GPRS, xDSL, or Wi-Fi, to name just a few examples.

It should be noted that the CDMA standard bodies (officially referred to as 3GPP2) have adopted the 3GPP Release 5 IMS standards as the baseline for their IMS standard, which they refer to as Multimedia Domain (MMD). However, due to having two separate working bodies with dissimilar access technologies and business drivers, some differences in the network architectures have appeared. There is an agreement between 3GPP and 3GPP2 to keep the two standards in agreement, although not necessarily in sync. Verizon Wireless™ in concert with several of its vendors have announced an initiative referred to as Advances to IMS (A-IMS), which is intended to bridge the gap between the two standards groups.

2.4 The Tenets of IMS

As was laid out in the previous sections, we saw through the evolutionary steps of the centralized architecture of the first-generation MSC to the highly distributed architecture of the third-generation IMS topology that a greater optimization of resources was obtained. Breaking the monolithic MSC into its fundamental elements and defining standardized interfaces between those elements produces greater optimization in design, development, testing, and manufacturing by network equipment vendors. This in turn has led to a greatly reduced time period in producing a product for market. Vendors' development schedules are now measured in months instead of years. And with this greater optimization and a migration to IP technology has come a reduction in capital costs to the operator.

For the rest of Part I, we will look at each of the IMS planes and their individual elements. We will examine how the 3GPP standards bodies have defined each element plus a discussion on trends in how the vendor community has implemented the standards. However, first we will start with a discussion in Chapter 3 on how the standards bodies that impact IMS are organized, and then follow with a look at each of the IMS components. But before we close this chapter, we want to quickly summarize for easy reference the key tenets of IMS, as indicated in the 3GPP IMS Stage 1 document, which have directed how the architecture and standard has been defined and how it should evolve in subsequent releases.

- First and foremost, the entire underlying infrastructure for telecommunications undergoes a fundamental shift as networks move from silo circuit switch voice application networks overlaid with a packet-switched data application network to a *single packet-switched network for both voice and data applications.*
- With IP as the base technology, *the Internet world and its paradigm can now be applied to the telecommunications industry.* This new paradigm brings with it the concepts of rich voice (voice service combined with other real-time applications) and multitasking (the ability to run multiple applications simultaneously).
- *A single common network infrastructure that is independent of access technology.* Thus, the same application can be used to offer a service to an end customer regardless of their current access technology (e.g., GSM, UMTS, WCDMA, Wi-Fi, DSL, etc.). This same service can also follow the end customer as he roams between access technologies (e.g., wireless/wireline convergence).
- *End-to-end quality of service (QoS) appropriate to the application.* The best-effort approach of the public Internet is not reliable enough to offer real-time applications performance requirement. Anyone who has been on the other end of a VoIP call using the public Internet understands the jitter and latency problems when a best-effort QoS approach is used. Thus, QoS along with policy control can enable a differentiated service offering.
- As the legacy circuit-switched network migrates toward an all IP-based network with its many distributed elements, the ability to effectively manage these new network resources becomes more complicated. *An effective policy control framework must be put in place to administer, manage, and control access to network resources.*
- A *Virtual Home Environment (VHE) allows for the customer's service environment to be portable across network boundaries and between terminals.* The concept of a VHE is such that users are consistently presented with the same personalized features, user interface customization, and services in whatever network and whatever terminal (within the capabilities of the terminal and the network) wherever the user may be located.

■ Since an IMS network most likely will not go into a greenfield environment, it must *support the ability to interwork with existing networks such as the PSTN, ISDN, legacy PLMN, and Internet users.* Since IMS roll-out to the end customer is expected to be a gradual event and not a flash cut, IMS users must be able to communicate with non-IMS users.

■ The security of the IMS is a foundational tenet. As laid out in 3GPP TS 33.120, *"3G security will build on and improve the security of the second-generation systems."* Both access domain security (between terminal device and the network) and network domain security (intra- and internetwork) are addressed. IMS introduces a new application on the 3G SIM card (aka, Universal Integrated Circuit Card or UICC) known as ISIM (or IMS Subscriber Identity Module). It provides security at the IMS layer and is supplemented by transport layer security (such as the radio access layer in UMTS that implements its own security mechanisms unique to that layer).

■ The ability to charge for services provided is essential for the existence of any service provider's network. However, as operators very well know, they must be able to offer flexible charging models that are responsive to the needs of the marketplace. *Charging models for the IMS must be flexible to allow multiple different charging models to address a competitive marketplace as well as to address the regulatory environment and numerous internetwork conditions.* Examples of the charging flexibility accounted for in the 3GPP specifications include prepaid, postpaid, calling party pays, and called party pays.

Endnotes

1. Or more typically, a new switch was required when an operator reached ~80% of the rated switch capacity as a safety margin.
2. Note that even though both TIA and GSM defined the Authentication Center (AuC) as a separate entity, most implementations had the AuC co-located with the HLR.
3. In the classic Intelligent Network (IN) architecture, the gsmSCF would be the service control point (SCP).
4. 3GPP R99 actually defined the first implementation of UMTS but the work in R99 focused on the RAN; no specifications were defined in R99 for support of a 3G core network.

Further Reading

3GPP TR 22.121, "The Virtual Home Environment," Section 4.
3GPP TS 23.205, "Bearer-Independent Circuit-Switched Core Network; Stage 2."
3GPP TS 23.228, "Service Requirements for the Internet Protocol (IP) Multimedia Core Network Subsystem; Stage 1."

3GPP TR 32.200, "Telecommunications Management; Charging Management; Charging Principles," September 2005.

RFC 3435, "Media Gateway Control Protocol (MGCP)," Andreasen, F. and Foster, B., January 2003.

Chapter 3

Standards Organizations for the IMS

There are hundreds of standards development organizations (SDOs), some well known to the public and others only known in their relative areas. They are all created for a purpose, and each has its focus area. IP Multimedia Subsystem (IMS) started as the telecommunications standards to bring the Internet Protocol (IP) into a traditional telecommunications environment that is based on circuit switch technologies. When it comes to IMS-related standards development, the three SDOs discussed in this chapter play the most important roles and provide the biggest impacts.

3.1 Third-Generation Partnership Project (3GPP™)

3GPP was established in 1998 by the collaboration agreements of a number of telecommunications standards bodies that are known as Organizational Partners (OP). The current OP of 3GPP are the following:

- **ARIB**: Association of Radio Industries and Business (ARIB) is a Japanese agency that develops standards to promote the efficient use of the radio spectrum and designated frequency change support.
- **CCSA**: China Communications Standards Association (CCSA) is a nonprofit organization in China, approved by the governmental Ministry of Information Industry, established by enterprises and institutes to carry out standardization activities in the field of information and communications technology.

- **ETSI**: European Telecommunications Standards Institute (ETSI) is a non-profit organization recognized by the European Commission as a European Standards Organization. It develops standards for information and communications technologies that are applicable to both wireless and fixed-line access technologies.
- **ATIS**: Alliance for Telecommunications Industry Solutions (ATIS) is accredited by American National Standards Institute (ANSI) to develop standards and solutions for the North American communications markets.
- **TTA**: Telecommunications Technology Association (TTA) is a telecommunications standards development organization in South Korea.
- **TTC**: Telecommunications Technology Committee (TTC) is an organization authorized by Japan's Ministry of Internal Affairs and Communications to develop and promote standards for telecommunications in Japan.

3GPP OPs provide the representations from the major SDOs in Europe, North America, and Asia; the standards specifications it develops are widely used by equipment vendors, network service providers, and applications developers around the world, and provide the biggest impact on the evolution of today's telecommunications industry.

The original goal of 3GPP was to develop technical specifications and technical reports for the third-generation mobile system based on the evolved Global System for Mobile Communications (GSM)[1] technology. Since then, the scope of 3GPP has been expanded to take over the maintenance of existing GSM technical specifications and technical reports and its technology evolution.[2]

3GPP consists of the project coordination group (PCG) and technical specification groups (TSGs) as illustrated in Figure 3.1.

The project coordination group is an administration and management body that is responsible for overall schedule coordination and management of technical work activities to ensure a timely development of a particular specifications release.

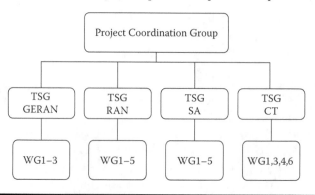

Figure 3.1 3GPP organizational structure.

The specifications are developed in TSGs. There are currently four TSGs in 3GPP; each of the TSGs creates its own working groups (WGs) to develop the relevant technical specifications or technical reports. The following subsections give a more detailed description of these four TSGs.

3.1.1 TSG Service and System Aspect (SA)

TSG SA is responsible for the overall architecture and service capabilities specifications development. There are five working groups under TSG SA to develop specifications on service requirements (SA WG1), system and service architecture (SA WG2), security (SA WG3), codec (SA WG4), and charging and network management (SA WG5). A Work Item Description (WID) is used to start a new feature or capability development, and they are normally started in SA WG1 (SA1). Once the work item has been agreed to in SA1 and approved by SA, a relevant work item will be introduced to other working groups depending on the nature of the original work item.

3.1.2 TSG Core Network and Terminals (CT)

TSG CT is responsible to develop detailed technical specifications[3] regarding terminal interfaces, capabilities, and core network aspect. TSG CT is the result of the merger of two earlier TSGs, TSG Core Network (TSG-CN) and TSG Terminals (TSG-T), in 2005. There are four working groups[4] under TSG CT to develop the relevant technical specifications and technical reports.

3.1.3 TSG Radio Access Network (RAN)

TSG RAN is responsible for the development of the functions, requirements, and interfaces of the Universal Terrestrial Radio Access (UTRA) network, the so-called third-generation Universal Mobile Telecommunications System (UMTS™) network. There are five working groups under TSG RAN to develop the relevant radio technical specifications and technical reports.

3.1.4 TSG GSM/EDGE Radio Access Network (GERAN)

TSG GERAN is responsible to develop technical specifications and technical reports for the second-generation wireless network, that is, GSM network. There are five working groups under TSG GERAN to maintain and develop relevant specifications.

3.1.5 IMS Specifications Development

IMS was developed by 3GPP and later adopted by 3GPP2.[5] The International Telecommunications Union[6] Telecommunication Standards Sector (ITU-T) Next Generation Networks (NGN) Global Standards Initiative (GSI)[7] approved IMS for NGN in November 2006.

The IMS development was started in TSG SA1 back in 1999 as a study item and was baseline with the completion of TS 22.228 version 1 in September 2000. TS22.228 is the stage 1 requirements document leading toward the development of a series of IMS technical specifications. The IMS standards are developed mainly in TSG SA and TSG CT. The first IMS deployment is based on IMS version 5 specifications, which were started in late 2000 as the continuation of version 1 development; its development was completed in 2006. Currently, 3GPP is developing IMS version 8 specifications that are targeted to be completed in late 2009.

3GPP is a membership-oriented organization; only registered members have the right to participate in the specifications development. The membership is composed of companies and organizations. TSGs have the authority to approve work items and specifications; they normally meet four times a year. The working groups under TSGs set their own meeting agenda and schedule to complete their tasks (i.e., work items). Face-to-face meetings are the main means to advance work. When it comes to conflict resolution, 3GPP uses the majority-rule procedure where each member company or organization has equal voting rights. For a complete feature development such as IMS, 3GPP will develop stage 1 (requirements) specifications, stage 2 (architecture solution) specifications, and stage 3 (call flows and procedures) specifications.

When 3GPP was developing Wideband Code Division Multiple Access (WCDMA)-based UMTS IMS, 3GPP2 was developing a similar solution to converge telecommunications and IP networks for the narrow-band CDMA systems. After a long investigation and discussions, 3GPP2 adopted[8] the IMS developed by 3GPP to avoid duplicating similar efforts and to also avoid potential incompatible solutions. 3GPP2 named its adopted 3GPP IMS the Multimedia Domain (MMD).

3.2 Open Mobile Alliance (OMA)

OMA is the most important of the international SDOs working on developing mobile service enablers[9] based on the 3GPP IMS Technical Specifications. OMA maintains a vision of openness to develop technical specifications for broad industry participation and adoption. It has established an openness policy ensuring public availability of its specifications.

OMA was created in June 2002 as the integration of the following industry fora:

- *WAP Forum:* Focused on standards related to mobile information and telephony services
- *Wireless Village:* Focused on instant messaging and presence development for mobile system
- *SyncML Initiative:* Focused on data synchronization
- *Location Interoperability Forum:* Focused on mobile-location-related application development

- *Mobile Games Interoperability Forum:* Focused on mobile gaming service development
- *Mobile Wireless Internet Forum:* Focused on bridging the gaps between mobile and Internet technologies for 3G wireless services

All these fora were developing applications and protocols for mobile services. It was a big burden for companies[10] in the mobile industry to cover all activities of these fora; different meeting schedules, different governance rules, different decision-making procedures and different priorities in applications development, etc. OMA's creation was the answer to solve all these issues. After the creation of OMA, all of the specifications and existing development activities in these fora were carried into OMA and continued in relevant working groups. As an example, the Instant Messaging and Presence Service (IMPS[11]) development in Wireless Village was continued in the OMA Messaging Working Group (under the IM sub-working group) and finished there.

One main goal of OMA is to develop open technical specifications based upon market requirements, which provide consistent solutions for mobile applications and services to reduce industry implementation efforts. OMA published its technical specifications as service enablers for vendors to develop their products. The OMA organizational structure is illustrated in Figure 3.2.

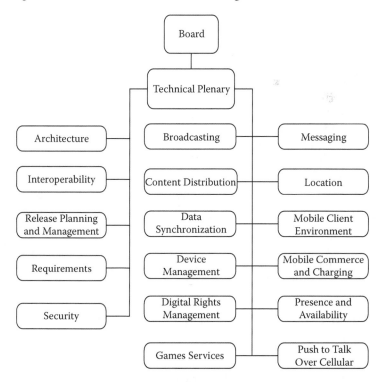

Figure 3.2 OMA organizational structure.

The Board is the governing body of OMA which manages budgets and all activities. It created the Technical Plenary (TP) as a committee to be responsible for all technical activities including service enabler's development. TP creates working groups to manage technical specifications development activities. A working group is closed by TP when it has accomplished its roles in technical specifications development. As shown in Figure 3.2, there are currently 17 working groups in OMA. The working groups that are relevant to IMS-based service enablers developments are Requirements, Architecture, Messaging, Push-to-Talk over Cellular (PoC), and Presence and Availability Group (PAG).

Similar to 3GPP, OMA uses the three-stage specifications development process. A WID is used to start an Enabler development.[12] Once a work item is approved, stage 1 (requirements) specification, stage 2 (architecture solution) specification, and stage 3 (call flows and procedures) specifications are developed. In addition to these technical specifications, Enabler Test Requirements (ETR) and Enabler Test Specifications (ETS) are developed in the Interoperability Working Group. A service enabler development is not completed until it has gone through interoperability (IOP) testing. OMA holds several test fests a year to accomplish this.

A minimum of four cosigned companies are needed to start a new work item.[13] The new work item proposal normally is presented to the relevant working groups to seek feedback, revisions, and support[14] before asking for approval. Only TP has the authority to approve a work item and assign it to the appropriate working groups where the development starts.

An OMA enabler development goes through three stages; it is in the *draft stage* when it is being developed. Once the development is completed[15] it is promoted to the *candidate stage*. It will only become an *approved enabler* after the completion of IOP tests and approved by TP. The IOP test is used to validate the *candidate specifications*. An approved enabler is the OMA standards.

Similar to 3GPP, OMA uses face-to-face meetings as its main activities to advance the specification development. Currently, the working groups meet face-to-face six times a year; TP face-to-face meetings are colocated with working group meetings and occur four times a year. Many working groups schedule interim meetings between regular meetings to further progress their works. Conference calls are used frequently to progress the specification development; however they are typically not as effective as face-to-face meetings.[16] E-mail lists are used mainly for offline discussions and timely approval decisions.

3.3 Internet Engineering Task Force (IETF)

IETF is an open international community of network designers, vendors, and researchers concerned with the evolution of the Internet architecture and the smooth operation of the Internet. It is best known for developing all the Internet protocols (IPs) that are used today over the Internet. The first IETF meeting was

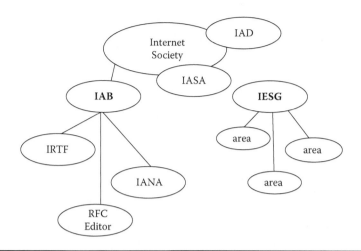

Figure 3.3   IETF organizational structure.

held in January 1986 with only 21 attendees; now there are more than 1000 participants attending IETF meetings. Figure 3.3 illustrates a high-level IETF organizational structure.

The Internet Society is a nonprofit organization that has been the home base for IETF staff since 2005. Its Internet administrative director (IAD) oversees IETF meetings and provides operational supports. The well-known IETF consists of two main organizations: the Internet Engineering Steering Group (IESG) and Internet Architecture Board (IAB).

The IAB is responsible for long-term evolution of the Internet and review of the creation of new IETF working groups. The IAB sponsors the Internet Research Task Force (IRTF) that provides in-depth reviews of specific Internet architecture. The Internet Assigned Numbers Authority (IANA) is responsible for the global coordination of the Domain Name Service (DNS)[17] Root, IP addressing,[18] and other IP resources.[19]

The IESG is responsible for technical management of IETF activities and Internet standards development process. IESG has the authority to approve the publication of IETF specifications. IESG creates *areas* to oversee the technical activities and specifications development. There are currently eight areas in IETF:

- *Application Area:* Responsible for Internet protocols seen by user programs such as e-mail and the Web
- *General Area:* A catch-all area for working groups that don't fit in any of the defined areas
- *Internet Area:* Responsible for protocols that move IP packets and DNS information

- *Operations and Management Area:* Responsible for operational aspects, network monitoring, and configuration
- *Real-Time Applications and Infrastructure Area:* Responsible for delay-sensitive interpersonal communications
- *Routing Area:* Responsible for solutions on getting packets to their destinations
- *Security Area:* Responsible for authentication and privacy
- *Transport Area:* Responsible for special services for special packets

There are working groups created under each area to develop technical specifications. Currently, there are about 115 active working groups in IETF; SIP, SIPPING, SIMPLE, MMUSIC, and XCON working groups are relevant to IMS standards development.

Different from 3GPP and OMA working methods, IETF working groups conduct their specification development activities mainly through e-mail discussions. In a sense, an IETF working group is just a mailing list with a little bit of moderation from the working group chairs. Anyone could "join" a mailing list and participate in its group discussions. It demonstrates the true "openness" in the standards development process of IETF.

Unlike 3GPP and OMA, there is no membership in the IETF.[20] Anyone can join a working group discussion by sending a request to be added to that working group's mailing list. IETF face-to-face meetings occur three times a year. For a week-long, face-to-face meeting to accommodate 100+ working groups,[21] very little time could be allocated to a working group. Most of the working groups use face-to-face meeting to solve controversial or critical issues and socialize new ideas.[22] Additional face-to-face meeting could be arranged by working group chairs if there is a need to advance its works.

The specification published by IETF is called the Request for Comments (RFC). An RFC may be published as *informational, experimental, standards track,* or *best current practice.* An RFC always starts as an Internet draft document and anyone can submit an individual Internet draft. An individual Internet draft is adopted as a working group draft once approved by that working group. A working group draft will be reviewed and tracked by the working group and hence is believed to be more "thorough." Once an RFC is published, it will never be revised. A revision of an existing RFC is published as a new RFC with an indication as to which existing RFC it replaces.

The informational and experimental RFCs are not meant to be Internet standards. However these two kinds of specifications could provide useful information related to implementing the standards-tracked specifications. Very few RFCs are actually on the standards tracks, which require going through rigid reviews, implementations, and interoperability testing. IESG is the organization that has the authority to approve an Internet draft to be on the standards track. The approved standards track RFC is first published as a *Proposed Standard.* After a minimum six month public review and with at least two independent implementations and interoperability tests, the RFC author(s) can request that their Proposed Standard be promoted to a *Draft Standard.* In rare situations, the IESG will promote a Draft

Standard to an *Internet Standard* (also known as a *Full Standard*) when there is evidence of widespread deployment. The Full Standards are normally reserved for protocols that are absolutely required for the Internet to function, and there are very few of them today.[23] One of the criteria to promote a Draft Standard to an Internet Standard is that it has been widely deployed and used in the Internet. In fact, many of the Internet Protocol standards we use today are Proposed Standards.

IETF culture is very different from a regular SDO such as 3GPP or OMA; it is obvious just from its dress code to meetings. The attendees of IETF meetings tend to be more "liberal"; you will see many participants dressed in shorts and sandals in summer meetings. Every meeting attendee has an equal right in voting; no companies or organizations are recognized. Its liberal culture also is shown in its voting process: instead of a formal show of hands to count votes, most of the working groups use "hum" to count votes. The louder the hum, the more votes it represents.

3.4 Conclusion

3GPP gave birth to the IMS concept in 1999 and started its initial framework development in 2000. Its development on IMS standards has never stopped and is now finishing work on Version (Release) 8[24] of the specifications development and starting work on Version 9. It has enriched the IMS technical specifications to be adapted outside of 3GPP. OMA covers the service aspect of IMS standards development and has developed several service enablers that could be used to offer IMS-based services. It continues to identify possible IMS-based services and develop the specific enablers. IETF is the birthplace of Session Initiation Protocol (SIP) and continues to enrich the relevant SIP protocols that will be used in IMS environment. These three SDOs have been working closely[25] together to coordinate SIP-related protocol development to make IMS-based services commercialization possible.

There are other SDOs and industry fora also developing IMS-related services and implementation guidelines. GSMA has developed "IMS Roaming and Interworking Guidelines (PRD[26] IR.65)" to guide its members on how to interconnect with IMS. ATIS, ETSI, CableLab, OMTP,[27] and the IMS Forum are all developing IMS-related guidelines and technical specifications, and play a role in deploying IMS-based services.

Endnotes

1. Global System for Mobile Communications (GSM) is the most popular and widely available mobile phone standards in the world.

2. GSM is viewed as a second-generation mobile communications system. 3GPP standards are an evolutionary path for the GSM systems to migrate to 3G. Advanced Mobile Phone System (AMPS) is the first-generation mobile communications system developed by AT&T™ Bell Labs and was first introduced in North America in 1983.

3. 3GPP uses the three-stage specification development process to develop a feature; stage 1 is the requirements development, stage 2 is the architecture solution development, and stage 3 is the detailed specification development.

4. IMS requirements and service architecture were developed in TSG-SA and are most relevant to the scope of this book, so their sub-working groups are mentioned in this chapter. TSG-CT develops the detailed specifications for IMS that are used mainly by the vendor community to develop their IMS products and are outside the scope of this book. TSG-RAN and TSG-GERAN have minor or no roles in IMS specifications development. Refer to the "3GPP Overview" reference in the "Further Reading" section for their sub-working groups' organizations.

5. Third Generation Partnership Program Two (3GPP2) was formed one day after the formation of 3GPP and could only use 3GPP2 as the name of its partnership.

6. ITU is the leading United Nations agency for information and communication technologies.

7. ITU-T NGN Global Standards Initiative (GSI) focuses on developing the detailed standards necessary for NGN development to give the service providers the means to offer the wide range of services expected in NGN. NGN-GSI harmonizes different approaches to NGN architecture worldwide.

8. With so many companies covering both 3GPP and 3GPP2 activities, it makes sense to harmonize the IMS solutions between these two SDOs.

9. OMA defines an enabler as "A technology intended for use in the development, deployment or operation of a service; defined in a specification, or group of specifications, published as a package by OMA." It should be added that this "technology" can be shared across multiple services for a variety of reasons as discussed in Chapter 1.

10. These companies include mobile network operators, network equipment vendors, mobile device vendors, and software developers.

11. IMPS is the Instant Messaging service developed in Wireless Village and later completed in OMA. OMA Messaging Working Group also developed another IM Enabler based on 3GPP IMS and IETF SIP/SIMPLE protocols. OMA stopped further IMPS development after completing version 1.3 in favor of the SIMPLE-based IM development.

12. A work item normally goes through socialization, amendment, and approval processes.

13. Depends on the nature of the work item; it could lead to the development of a service enabler, a reference release, or a white paper. A reference release or a white paper is not an OMA standards.

14. This is called *work item socialization*.

15. OMA has a review process to make sure a specification development is completed and approved by member companies.

16. A big challenge to scheduling an OMA conference call is to find a time that is convenient to all of its members located in different time zones around the world. Conference bridge quality and moderation of a big group in a conference call are also challenges.

17. DNS provides mapping from human-readable computer hostnames into an IP routable address. The DNS root is the uppermost part of the DNS hierarchy, also known as *top-level domain*; ".com," ".org," ".net," etc., are widely known DNS roots.

18. IANA coordinates the Internet Protocol addressing systems as well as the Autonomous System Numbers (ASN) used for routing Internet traffic. The IP addresses are assigned by local registry authority.
19. IANA assigns codes and numbers used in a variety of Internet protocols.
20. Many claim that IETF is the only SDO that provides TRUE openness because of its no-membership, no-fee, individual-based participation rule.
21. Not all working groups meet in IETF face-to-face meetings; actually most of them don't meet face-to-face at all!
22. Both IETF and OMA use "bird of feather (BoF)" meetings to socialize new ideas.
23. As an example UDP (RFC 0768) is an Internet Standard. There are only 76 Standard RFCs out of 5280 RFCs today.
24. Published or draft specifications are identified by their version even though the common vernacular is to call them *releases*.
25. Both 3GPP and OMA have official liaisons with IETF.
26. Permanent Reference Document (PRD) is the official document published by GSMA.
27. Open Mobile Terminal Platform (OMTP) is a forum created by mobile operators and supported by vendors to work on standardization of mobile device requirements. It is developing IMS functional requirements for device platforms.

Further Reading

3GPP Overview, http://www.3gpp.org.
OMA Overview, http://www.openmobilealliance.org.
IETF Overview, http://www.ietf.org.
Permanent Reference Document IR.65, "IMS Roaming and Interworking Guidelines," GSMA http://gsma.org.

Chapter 4

The IMS Control Plane

The IMS Control Plane[1] is where the necessary message exchange takes place for the registration of a user onto the IMS network and for the establishment of a session. In a pre-IMS network, the establishment of a session would be synonymous with the establishment of a call (or call connection). With IMS, as all services involve Internet Protocol (IP)-based data (including traditional voice calls, e.g., VoIP), a user establishes a data session (or more simply a session) in order to invoke any kind of IMS service.

The elements that make up the Control Plane are

- Call Session Control Function (CSCF), which consists of three components:
  - Proxy CSCF
  - Interrogating CSCF
  - Serving CSCF
- Home Subscriber Server (HSS), which would also include the Subscription Locator Function (SLF) if there are multiple HSSs
- Breakout Gateway Control Function (BGCF)
- Signaling Gateway (SGW)

There are two more elements that make up the Control Plane: Media Gateway Control Function (MGCF) and the Media Resource Function Controller (MRFC). Because they are closely tied with the Bearer Plane elements, IMS-Media Gateway (IMS-MGW) and Media Resource Function Controller (MRFC), respectively, they will be discussed in the next chapter.

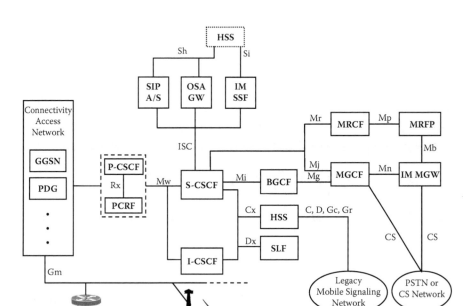

Figure 4.1   IMS reference architecture.

Figure 4.1 shows the reference architecture as defined in 3GPP TS 23.002. We will use this figure throughout this and the following two chapters to show how the different elements in the IMS relate to one another.

4.1 Call Session Control Function (CSCF)

The CSCF provides call processing and control functions analogous to those provided by a circuit switch. In addition, since IMS is an IP-based network, the CSCF provides the session control function. The standards have defined the CSCF as a SIP server that consists of three distinct functional components: Proxy, Interrogating, and Serving. Most early implementations combine all three components into a single product for performance reasons (implementing an internal proprietary protocol between the different CSCFs is more efficient than SIP) and economic reasons (it is less expensive for a customer to purchase and operate one box rather than three). As IMS networks mature, the Proxy CSCF component is expected to become its own distinct element. Only in the largest network deployments is a physically separate Interrogating CSCF component likely to make sense.

4.1.1 Proxy CSCF (P-CSCF)

The P-CSCF is the initial entry point into the IMS although it is not necessarily the initial entry point for the end user. An end user must connect through some type of access network prior to connecting to an IMS network. Examples include the GPRS network, which consists of the Gateway GPRS Support Node (GGSN) and the Serving GPRS Support Node (SGSN) for users connecting from a cellular network or a Packet Data Gateway (PDG) for a Wi-Fi network (Figure 4.2). A discussion of these elements is beyond the scope of this book. However, the reader should be aware that in order for IMS to be an access-independent network, there must be access-dependent network elements that are required to interconnect to the IMS entry point (i.e., P-CSCF).

The user equipment (UE) or terminal device sends all IMS signaling messages to the P-CSCF. There can be multiple P-CSCFs in the user's home network through which the UE would connect. Alternatively, the UE may connect to a visited network's P-CSCF while roaming, as the standards support both implementation configurations. The UE is provided the address of its assigned P-CSCF during the registration process with its connectivity access network. If the UE is not provided this information, then the connectivity access network will provide a discovery mechanism using Dynamic Host Configuration Protocol (DHCP) to

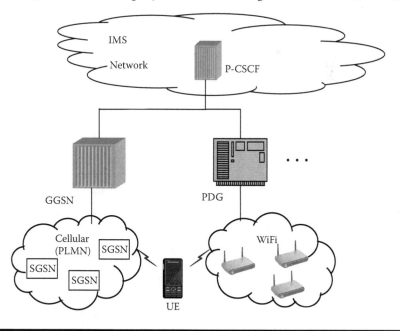

Figure 4.2 Different access points into the IMS.

provide the P-CSCF domain name and the address of a domain name server (DNS) to translate the domain name to an IP address.

For cellular networks, the P-CSCF is in the same network as the GGSN serving the subscriber. Connecting through the visited network's P-CSCF leads to some traffic routing efficiencies if the UE is trying to connect to another user in that same visited network (which is why such connections were included in the standards). However, most operators in their early IMS implementation will support only the home network P-CSCF configuration primarily for network simplification due to the immaturity of IMS roaming agreements and because of GGSN implementations being deployed within the home network environment only.

It should be noted that the P-CSCF may also incorporate the Policy and Charging Rules Function (PCRF) as a logical entity. The PCRF among other things would establish the quality of service (QoS) for a particular session. If the P-CSCF is implemented as a separate node (this is the most likely implementation scenario for operators implementing a centralized policy management strategy), then it would query the PCRF over the Rx interface for session establishment information.

The functions that the P-CSCF are to perform include the following:

- Receive all incoming SIP messages from the UE; processing and forwarding valid messages into the IMS and discarding invalid messages
- Deliver outgoing SIP messages to the UE from the IMS
- Handling of emergency call/session establishment requests
- Creating appropriate call detail records
- Perform SIP message compression/decompression as required
- Maintain a security association with the UE (see Chapter 7, Section 7.1)
- Authorizing QoS and bearer resources

4.1.2 Interrogating CSCF (I-CSCF)

The I-CSCF receives the SIP message forwarded from the P-CSCF and is instrumental in determining the S-CSCF that the UE will be assigned to. Upon receiving a REGISTER message from a UE, it queries the subscriber's HSS via the Cx interface to determine if the UE has already been assigned an S-CSCF or if not, what are the required capabilities to meet the subscriber's services needs or the operator's particular preferences for a subscriber. Knowing the needed capabilities for a subscriber will become more important as deployed networks become more mature.

Besides assigning the S-CSCF to the UE, the other primary role for the I-CSCF is to hide the internal workings of an operator's network from those interacting with it. In this role, referred to as the Topology Hiding Inter-network Gateway (THIG), the I-CSCF hides the topology, configuration, and capacity of an operator's network from the view of peer networks by changing or encrypting sensitive information such as domain names or the internal routing path of a message.

4.1.3 Serving CSCF (S-CSCF)

The S-CSCF is at the heart of the IMS. It performs the call processing function analogous to the call processing functionality that is the heart of the circuit switch, in addition to all session processing. All SIP messages that flow from or to the UE are routed through the S-CSCF. Like its circuit switch predecessor, the S-CSCF maintain all UE sessions state information, handles interactions with service platforms, handles requests for bearer resources (including network-based tones and announcements), and invokes charging functions. Depending on the situation, it acts as either a SIP proxy server or a SIP registrar.[2] As a SIP proxy server, it routes SIP messages to their appropriate destination (e.g., SIP AS). As a SIP registrar, it maintains a binding with the UE by taking the information in the REGISTER message and placing it (as appropriate) in the HSS.

Additional primary functions that the S-CSCF may perform (depending on the incoming SIP message) are as follows:

- Authenticating the UE (which includes obtaining the authentication vectors from the HSS) by means of IMS AKA (see Chapter 7, Section 7.2 for details).
- Upon registration, retrieving subscriber profile information, which would include service-related information such as application server selection priority.
- Translating an E.164 number to a SIP uniform resource identifier (URI) since routing within the IMS is based on SIP URIs.
- Monitoring and enforcing UE registration timers and maintaining session timers (which allow the S-CSCF to free up valuable resources used by hung or unsuccessful sessions).
- Routing originated calls to the appropriate destination endpoint. This routing could be to another S-CSCF for a call within the same IMS network, to another operator's I-CSCF for call completion to another IMS network, or to the BGCF for call routing to the PSTN or a circuit-switched network.
- Checking media resource requests within a SIP message. This involves checking the requests made in the Session Description Protocol (SDP) payload which is part of the SIP message (see Chapter 12 for a more detailed explanation of SDP). If the requested SDP is not consistent with the subscriber profile (e.g., the subscriber is requesting a video call service but is not subscribed for any video call service) or operator's policies, then the S-CSCF would reject the session request.
- Sending call detail/accounting information as indicated in the subscriber profile to the Online Charging System (OCS) for prepay accounts, or sending it to the Charging Collection Function (CCF) for postpay accounts.

It is envisioned that as more S-CSCFs are deployed, they will not be equally equipped for economic reasons (e.g., a "gold-plated" S-CSCF for high-end customers, "bronze-plated" for mass market customers, and possibly a dedicated S-CSCF for enterprise customers).

4.2 Home Subscriber Server (HSS)

Although one may view the S-CSCF as the heart of the IMS, certainly the HSS would be the brains of the IMS. The HSS is the repository of all subscriber profile data and authentication information as well as all relevant temporary subscriber data such as current registration status and location. The HSS can be viewed as the IMS version (or evolution) of the Home Location Register (HLR) from a Global System for Mobile Communications (GSM) or Code Division Multiple Access (CDMA) network. It comprises the legacy HLR function plus additional functions related to providing IMS functionality (originally called the User Mobility Server[3]). As the evolved implementation of an HLR, the HSS would certainly perform all the functions of the legacy HLR, such as the authentication (AuC) function. In addition, it must be able to support the functionality that comes with an IP-based (our IMS) network. There are three reference points through which other IMS elements interact with the HSS. The Cx interface handles interaction with the I/S-CSCF. The Sh interface allows application servers to query the HSS for subscriber profile data. The Si interface supports queries from the CAMEL Service Environment (CSE). The Si interface is the least mature of these interfaces and the least likely of the three standard interfaces that one is likely to find in actual implementations. All three of these interfaces make use of the DIAMETER protocol (see Chapter 16 for more detail).

Figure 4.3 shows the logical layout of the following different HSS functions as described in 3GPP TS 23.002.

- **User Security Information Generation:** As part of the security process, the HSS will generate user-specific authentication, integrity, and ciphering data.
- **User Security Support:** The defined authentication procedures to access the operator's network such as IMS AKA (see Chapter 7 for more information) for the IM CN subsystem. This would involve storing the User Security Information generated and providing this information to the appropriate network elements (e.g., MSC/VLR, SGSN, CSCF).
- **Access Authorization:** Authorizes requests from network elements such as the MSC/VLR, SGSN, or CSCF for a user to access the network.
- **Identification Handling:** Each network domain has their own unique identifiers (MSISDN and IMSI for CS, MSISDN, IMSI, and IP addresses for PS, Private, and Public User Identifiers for IMS) in which the HSS handles the identifier relationship based upon the network domain that the user is currently accessing.

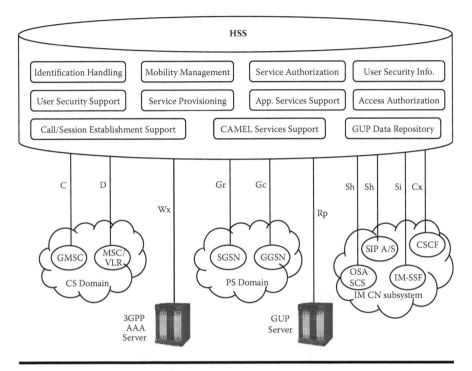

Figure 4.3 Logical functions for HSS.

■ Mobility Management: Manages the mobility of the user in the CS domain, PS domain, and IM Core Network subsystem.

■ Call/Session Establishment Support: The HSS supports the call/establishment procedures for the different domains according to the appropriate Third Generation Partnership Project (3GPP™) specification. It maintains information on the current S-CSCF that is serving the user and provides this information to appropriate network elements (e.g., the I-CSCF queries the HSS to determine if the user has already been assigned to an S-CSCF).

■ Service Provisioning Support: The HSS maintains and provides access to the service profile information of the user across all domains.

■ Service Authorization Support: The HSS updates the appropriate network element (e.g., MSC/VLR, SGSN, CSCF) with service information indicating whether the requesting user is authorized to invoke a particular service and if appropriate, what aspects of that service they are allowed to invoke (e.g., can place domestic calls but not international).

■ CAMEL Services Support: The HSS communicates through the IM-SSF to support CAMEL-based services in the IM CN subsystem. It contains information related to the user's CAMEL subscription.

■ Application Services Support: The HSS can provide user profile or data information to SIP Application Services or to third-party application servers through the OSA-SCS. As an actual implementation, many operators will maintain this type of information in a database separate from the HSS for security, performance, and economic reasons.

■ GUP (Generic User Profile) Data Repository: The HSS can provide access to the IMS user data through the Rp reference point. Note that the GUP function has not been universally embraced by the operator community.

As the central repository of all subscriber profile information, the HSS is shown in many architecture diagrams as a lone element (although it is typically mated for data redundancy reasons). Realistically, as a network's subscriber base grows, this growth may exceed the capacity of a single (physical) HSS, which leads to the likelihood of a network having multiple HSS elements. Other possible reasons for having more than one HSS in a network include an operator's desire to limit a network catastrophic outage or the acquisition of another carrier's network. When multiple HSS are present in the network, the I-CSCF and the S-CSCF need to know which HSS to query in order to get a particular user's subscriber profile information. Other elements such as the 3GPP AAA server or an application server typically need to obtain the identity of the HSS hosting a particular user's subscriber profile information. In these instances, standards have defined an element called the Subscription Locator Function (SLF) to provide information on which HSS maintains a particular user's profile information. The SLF maintains a mapping of each subscriber being served in the network to the particular HSS that stores the information about them. The mapping includes the address ID of the particular subscriber, which can take several forms such as a SIP URI, a TEL URI, or an MSISDN, along with the identification address of the subscriber's HSS. The HSS address can be an FQDN or an IP address.

The interface from the CSCF to the SLF is through the Dx reference point. Like the Cx reference point, messages exchanged through the Dx reference point use the DIAMETER protocol.

4.3 Breakout Gateway Control Function (BGCF)

Even though an IMS network offers many advantages over the legacy circuit-switched (CS) network, the CS network will still be around for decades to come, as different operators follow different transition plans based on their customer requirements and their own marketing plans. Because it is a requirement to have a global telecommunication system (i.e., one in which any phone can communicate with any other phone), the designers of IMS placed a node in the architecture to handle the interworking of the IMS domain with the PSTN/CS domain. This node, the Breakout Gateway Control Function (BGCF), gets its name because it is responsible for selecting where a breakout (from the IMS network) to the CS domain is to occur.

The BGCF receives the SIP message from the S-CSCF when the S-CSCF determines that it cannot continue routing the session through the IMS (typically because it cannot resolve addressing via DNS or ENUM/DNS). The BGCF will determine whether the breakout will occur in the same network or in another network. If it is to occur in the same network, the BGCF will select a Media Gateway Control Function (MGCF) to handle the interworking with the CS domain. If the breakout is to occur in another network, then the BGCF will forward the session signaling messages to a BGCF in that network. In early implementations of IMS, it will not be unusual for the breakout to occur toward a network that does not have a BGCF (no IMS deployment). In cases like these, the BGCF will hand off the call to its own CS network which in turn can complete the call using standard PSTN routing.

4.4 Signaling Gateway (SGW)

The Signaling Gateway (SGW) is not viewed as an IMS element although it is considered part of the control layer of an IMS network. This is because its function is below the SIP (application) layer. The SGW performs the transport layer signaling conversion function for control signaling messages to interconnect different signaling networks (i.e., an SCTP/IP-based IMS network with an SS7 [CS] network) while transparently passing the application layer protocols (e.g., ISUP, MAP, CAP, etc.). The SGW can be implemented as a standalone element, or it can be integrated with another element.

The two other control layer elements shown in Chapter 2, Figure 2.5 will be discussed separately (the SEG in Chapter 7 and the CCF in Chapter 10).

Endnotes

1. Note that another common industry term for the control plane is the *signaling plane*. IMS completes the separation of bearer and signaling plane control that was begun in the Softswitch/Media Gateway/Signaling Gateway era.
2. For further description of a SIP server and SIP registrar, see Chapter 10 (Overview of SIP) and RFC 3261.
3. The HSS was originally envisioned to consist of the HLR and the User Mobility Server (UMS) as described in 3GPP TR 23-992 where the UMS maintained the IP network service profile. Although it is no longer technically correct to describe the HSS to consist of the HLR and the UMS (as standards has expanded the definition beyond just IP network service profile), to the layman this is an easy way to think of the HSS. This is truer when looking at early wireless operators' implementations as they are tending to maintain their existing HLR and installing just the UMS portion of the HSS in their IMS network.

Further Reading

3GPP TR 23-992, "Architecture for an All-IP Network."

3GPP TS 23.002, "Network Architecture."

3GPP TS 23.228, "IP Multimedia Subsystem (IMS) Stage 2."

3GPP TS 29.228, "C_x and D_x Interfaces Based on the Diameter Protocol; Protocol Details."

3GPP TS 29.229, "IP Multimedia (IM) Subsystem C_x and D_x Interfaces."

RFC 3261, "SIP: Session Initiation Protocol," Rosenberg, J., Schulzrinne, H., Camarillo, G., Johnston, A., Peterson, J., Sparks, R., Handley, M., and E. Schooler, June 2002.

Chapter 5

The IMS Bearer Plane

As noted in Chapter 4, the Media Gateway Control Function (MGCF) and the Media Resource Function Controller (MRFC) inhabit the IMS Control (or Signaling) Plane. Because of its close relationship to the IMS-Media Gateway (IMS-MGW), which resides in the IMS bearer plane, the MGCF is covered in the present chapter. Similarly, the MRFC is covered in the present chapter to emphasize its relationship to the Multimedia Resource Function Controller Processor (MRFP).

5.1 IMS-MGW and Its Relationship to MGCF

Whenever a subscriber in a circuit-switched domain communicates with a subscriber in a packet-switched domain, conversion between circuit and packet bearers must take place. The IMS-MGW serves as the conversion point.

Figure 5.1 shows the positioning of the IMS-MGW with respect to its controller, the MGCF. The figure also situates the IMS-MGW and the MGCF with respect to the circuit-switched and the packet-switched (IP) domains. The protocol between the MGCF and the IMS-MGW (i.e., across the Mn reference point) is H.248. H.248 is a packet-switched protocol that runs over IP. Therefore, the Mn reference point is shown as inhabiting the IP domain. At one time, the Internet Engineering Task Force (IETF) and the Telecommunication Standardization Sector of the International Telecommunications Union (ITU-T) worked jointly on protocol development in this area. Some documents still refer to *Megaco*, which is the IETF name for H.248.

What kinds of instructions does the MGCF communicate to the IMS-MGW? Before answering this question, we need to look at the overall role of the MGCF

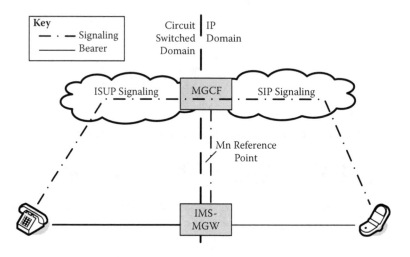

Figure 5.1   Placement of IMS-MGW astride circuit-switched and IP domains.

as depicted in Figure 5.1. Integrated Services Digital Network (ISDN) User Part (ISUP) is the predominant call control protocol for circuit-switched telephony. As Figure 5.1 suggests, the MGCF is responsible for conversion between ISUP and Session Initiation Protocol (SIP) call control signaling. In the process of performing this task, it establishes bindings between the following:

- Circuit-switched bearer plane identifiers
- IP bearer plane identifiers

The MGCF must communicate this binding information to IMS-MGW so that calls can seamlessly traverse the IMS-MGW. The IMS-MGW also needs to know the encoding scheme involved: there are many encoding schemes for voice as well as video sessions (Chapter 12 provides a list of these encoding schemes). IMS signaling determines the encoding scheme. Because it participates in SIP session control signaling, the MGCF knows which scheme is selected and can communicate that information to the IMS-MGW.

5.2 MRFP and Its Relationship to MRFC

Figure 5.2 shows the positioning of MRFP with respect to its controller, the MRFC. The protocol between the MRFC and MRFP (i.e., across the Mp reference point) is H.248. The protocol between the MRFC and the Call Session Control Function (CSCF) (i.e., across the Mr reference point) is SIP. The CSCF in the

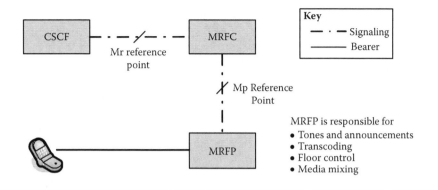

Figure 5.2   Placement of MRFC and MRFP.

figure can be Interrogating-CSCF (I-CSCF), Proxy-CSCF (P-CSCF), or Serving-CSCF (S-CSCF), depending on the use case.

Unlike the functional elements described in Section 5.1, the MRFC and MRFP do not interact directly with the circuit-switched domain. As shown in Figure 5.2, the MRFP may be called upon to perform a variety of tasks depending on the use case. We comment briefly on each type of task:

- *Tones and announcements:* The MRFP is responsible for playing tones and announcements. This can be something similar to "the number you have reached is no longer in service." It can also be something more amusing; perhaps, the calling party hears a recorded song to indicate that the called party's phone is ringing. The MRFP is responsible for playing the recording.
- *Transcoding:* As noted in Section 5.1, there are a number of voice- and video-encoding schemes. In IMS, encoding schemes are selected on the fly during session setup. So, there is no guarantee that the calling and called parties are using the same encoding scheme. As the name suggests, transcoding is the process of converting between encoding schemes.
- *Floor control:* Floor control is necessary in applications in which only one person is allowed to transmit at any given time. That person is said to "have the floor," and hence the name. The person who has the floor is the only person whom the other participants hear (or, in the case of certain video applications, the only person whom the other participants see). Push-to-Talk over Cellular (PoC) and Push to Video (PTV) are two examples of services that implement a floor control scheme.
- *Media mixing:* Media mixing is for conferencing applications. This is different from floor control. In the case of a voice conference, participants hear a mixture of sounds coming from the session participants. In the case of a video conference, the display may be divided among the participants.

Why are the four tasks just described grouped together? To a large degree, this is driven by technology. Specifically, all four tasks rely on digital signal processing hardware.

Further Reading

Recommendation H.248.1, "Media Gateway Control Protocol Version 2," ITU-T, 2002.
ISUP is specified by several standards organizations. For details, the reader is referred to the ITU-T Q.76x series, ANSI T1-113, and Telcordia GR-246 series.
3GPP TS 23.002, "Network Architecture."

Chapter 6

The IMS Application Plane

The Application Plane as one might suspect is the principal plane in which an operator would blend applications. Although this may be true, the Application Plane does not hold exclusive rights to this role. As we will discuss in Chapter 20, blending of applications can occur at several layers in the overall network as well as in several components.

Blending applications can be described as applications that have the ability (or capability) to interweave different (existing) applications, content sources, and devices to form a single higher-value service to the end user. These existing applications can be SIP based or, as discussed in Chapter 1, they can be applications from the Internet world. Typically, these Internet applications would be based on a Web services paradigm that would enter the operator's network through a Parlay X application programming interface (API). Other Internet applications may just require a straight Hypertext Transfer Protocol (HTTP) connection. Additionally, operators may desire to reuse their legacy suite of applications (e.g., Customized Application for Mobile network Enhanced Logic [CAMEL], Advanced Intelligent Networks [AINs]) for reuse in the IMS realm.

The network elements that make up the application layer typically interface with the Serving-Call Session Control Function (S-CSCF) and the Home Subscriber Server (HSS) as shown in Figure 6.1. The connection to the S-CSCF will be over the Internet Protocol (IP) Multimedia Subsystem (IMS) Service Control (ISC) Interface and over the Sh or Si interface for the HSS.[1]

The functional architecture for the Session Initiation Protocol (SIP) application server (AS), the IP Multimedia-Service Switching Function (IM-SSF), and the Open Services Access (OSA) Gateway is shown in Figure 6.2. In the figure, we see that Third Generation Partnership Project (3GPP™) introduces the concept of Service Capability Interaction Manager (SCIM) with very little explanation of its

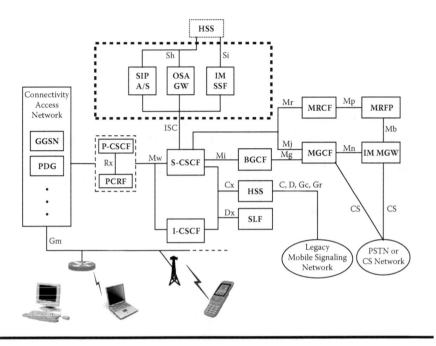

Figure 6.1   IMS reference architecture.

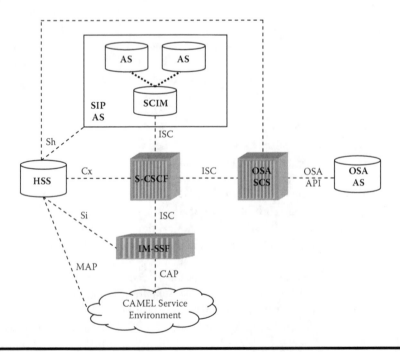

Figure 6.2   Functional architecture for application support.

role in the overall architecture. In fact, only a single sentence is provided in 3GPP TS 23.002 to explain its role: "The SCIM functionality is an application which performs the role of interaction management" [1]. The vagueness of this description allows for vendor or operator differentiation, not to mention the complexity of standardizing interaction management. In fact, when properly defined, the SCIM (or as we will later see—the service broker) will play an important role in blending applications. In Section 6.4, we will discuss in greater detail how these differentiations will be manifested and how the SCIM can be used to blend applications.

6.1 SIP Application Server

IMS applications will typically be implemented on a native SIP-based AS, although some AS vendors may choose to add a SIP interworking function to their existing non-SIP platform in order to minimize redevelopment efforts for an existing product. Alternatively, some may chose to deploy their application on a converged container, which is an AS that inherently supports different technologies (e.g., SIP, Web services). In each of these cases, the IMS network (and specifically, the S-CSCF) will communicate with the AS over the ISC Interface, which is a SIP-based protocol adhering to request for comments (RFC) 3261 and other related SIP RFCs. Because the design for IMS is to support sessions with multiple applications from multiple sources based on the user's subscription profile, a mechanism must be in place to select the appropriate AS for the user, and, in many cases when more than one service (i.e., AS) has been subscribed to, the order in which the ASs are to be invoked. When multiple ASs are listed, the response from the first AS can be used as input to the second AS (and so on) or to determine if there is even a need to invoke the next AS. The HSS stores, in the user's subscriber data profile, the list (or technically, a set of templates against which matches can be associated) of ASs and the order in which they are to be invoked in a list called the initial Filter Criteria (iFC), which is then shared with the user's S-CSCF during the registration process. The main content of the iFC passed from the HSS to the S-CSCF is as follows:

- AS address
- AS priority (invocation order)
- Default handling
- Subscribed media
- Trigger points
- Optional service information

The trigger point is an indication of when to invoke the iFC. In our example in Figure 6.3, Bob's phone is receiving (terminating) a call from John. The S-CSCF looks at the list (or template) provided by the HSS associated with an incoming call (or terminating trigger point) and goes out to the AS with address #5 (typically, this

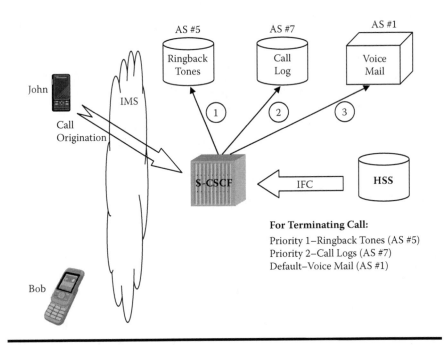

Figure 6.3 iFC example.

would be an IP address), where John hears Bob's ringback tone service play "White Christmas" (as it is near Christmas time in our example). Because Bob needs to have a record of all his calls for his job, the call information goes to a call logging service (AS #7) next. Finally, because Bob is occupied and cannot take John's call, he ignores the call and lets the default treatment take place, which is to route the call to his voice mailbox.

Note that the iFC order is static, and the priority order is always followed with no deviations. The service broker will be the mechanism used (as we will discuss in Section 6.4) to modify this order on a situational (dynamic) basis.

Our previous example showed the AS only reacting to a request from the S-CSCF. An AS can also be proactive and initiate requests into the network. They are able to establish a session leg to a user, or they may query a network enabler (such as the presence server discussed in Chapter 16). This is useful in cases such as third-party call control.

One last note about the iFC: even though all our discussions and examples were about SIP AS being in the iFC list, there is no reason why the iFC cannot contain a reference to an AS behind an OSA Gateway or a CAMEL service control point (SCP) behind an IM-SSF. In these cases, the OSA Gateway and the IM-SSF will appear as a SIP AS to the S-CSCF, as we will see in the following sections.

In addition to the ISC Interface, the SIP AS also supports the Sh interface, which is used to obtain user-related data stored in the HSS or transparent data also stored in the HSS. Transparent data is data used by any application but maintained in the HSS. The HSS has no knowledge about how it is used by the AS; it only knows its database location (thus, the data is transparent to the HSS). The application may require certain information about the user such as subscribed service data, current cell location, or currently visited network. Other data about a user's application may be stored transparently in the HSS. This transparent data is not understood by the HSS and the HSS is used only as a data repository.

The Sh interface may not be universally implemented by all operators. Many operators are concerned about the added traffic load and number of new network elements that would be accessing the HSS. Given the critical role that the HSS plays in keeping the network alive, operators will seek to reduce their risk of a network catastrophe. It is envisioned that large operators would implement a separate network database that would handle these "nonessential" queries for information. A side benefit is that these separate databases can be general-purpose databases, which offer a greater flexibility than current HSS implementations but with a trade-off of less reliability. This results in extra work and expense for the operator because of dual-provisioning requirements and maintaining a second database infrastructure; however, larger operators with sufficient staff can opt for this type of solution. In smaller networks, where the traffic load on the HSS is less and there is limited staff to support a second database infrastructure, housing the information on the HSS and providing access via the Sh interface may be a more appropriate solution.

6.2 IP Multimedia-Service Switching Function

The IP Multimedia Service Switching Function (IM-SSF) is designed to allow mobile operators to reuse their existing Global System for Mobile Communications (GSM) or 2G application infrastructure with their IMS network. This was done to avoid the high cost of redeveloping and redeploying these applications. The IM-SSF is specifically designed to allow IMS networks to reuse the CAMEL[2] suite of applications already deployed. As a practical example, many GSM operators have deployed their prepaid implementation using CAMEL technology. These implementations have been very expensive and carry a high operational burden, and in many cases it would be prohibitive to replicate and support a second prepaid network.

The IM-SSF is designed to take the SIP message from S-CSCF and translate it to an equivalent CAMEL message. Thus, the IM-SSF appears as a SIP AS to the S-CSCF and as a gsmSSF (GSM Service Switching Function) to the gsmSCF (GSM Service Control Function), which is a CAMEL AS. Besides translating messages between the IMS and CAMEL environments, the IM-SSF also maintains CAMEL network features to facilitate the interworking functions. These features include handling the GSM call state model and maintaining trigger detection points, service keys, and AS (gsmSCF) addressing. Like the SIP AS, the IM-SSF has a mechanism in the Si interface to interact with the HSS to retrieve user and service profile data.

One other item of note (or speculation) before we conclude our discussion on the IM-SSF. IMS standards were developed by 3GPP standards bodies, which have their origin in the GSM industry standards bodies; thus, it is no surprise that they would make sure there were mechanisms to interact with legacy implementations (i.e., CAMEL). There is no reason to believe that an element similar to IM-SSF could not be developed to support other Intelligent Networks (INs) similar to CAMEL for fixed (wireline) networks implementation of an IN such as AINs or for a Code Division Multiple Access (CDMA) Wireless Intelligent Network (WIN). As the 3GPP standards body continues its liaisons with other standards bodies (e.g., ATIS, 3GPP2, etc.) as well as mergers with organizations such as European Telecommunications Standards Institute (ETSI) having a traditional wireline history, it is reasonable to expect that other legacy networks will be supported.

6.3 Open Services Architecture Gateway

The Open Services Access (OSA) effort in 3GPP[3] is an adoption of the efforts of the Parlay Group.[4] Consequently, the terms OSA and Parlay are used (within telecom circles) to refer to the same capability, namely, an architecture that enables application developers to access and make use of an operator's abstracted network functionality through open and standardized APIs. OSA defines a set of service capability servers (SCSs), each of which is a function that abstracts a network element and provides an OSA (Parlay) Interface toward an application. These SCSs are typically deployed on a gateway platform that provides additional functionality such as security, service level agreements (SLAs), and charging and account management (which are gathered as a collection of capabilities in the Parlay/OSA framework).

In reality, an operator would never just deploy an OSA SCS as depicted in the 3GPP standards (Figure 6.2). Instead, an operator would deploy an OSA Gateway that contained the OSA SCS. The S-CSCF would then interface with the OSA SCS on the OSA Gateway over the ISC Interface. Thus, the OSA Gateway would have the appearance of a SIP AS to the S-CSCF (similar to that of IM-SSF appearance). Operators are beginning to deploy OSA Gateways as a means of allowing the third-party developers to offer their applications to the operator's customers. The first generation of third-party services deployed by operators saw a "walled garden" approach for numerous marketing, operational, and security reasons. Later implementations saw the usage of application certificates to address some of the operators' security and application integrity concerns. The OSA Gateway is designed to alleviate many of the operators' concerns and to fit into an overall framework to support the deployment of third-party services.

What this will allow is the opportunity for the world of Internet developers to access information about a particular user in order to provide new services that users can request. This is expected to increase the number of available applications to the end user by several orders of magnitude. The challenge for the operator will be to manage the sheer number of potential new applications; the developer will have the challenge of determining how to market their service in a sea of new services. Some industry observers argue for a completely open (a.k.a., "wild west") environment to download any application to any device, much as is done today with a user's PC. Although this may be an acceptable model in the PC environment, care must be taken not to naively think this model would also apply to the mobile device world. As any novice PC user knows, too many applications opened or running in the background will greatly slow down the performance of their machine. In the case of a mobile device (which typically has a much smaller processor and memory size) this could mean the difference between the mobile device receiving or missing an incoming call. As the collective ecosystem grows, the opportunities presented by this growth will need to be balanced with the management of security, integrity, performance, and overall service quality.

There is one last item regarding the OSA Gateway and the third-party developer to discuss. The Internet application ecosystem has spawned millions of developers who are primarily using WSML (Web Services Mark-up Language) for developing applications. Parlay was originally designed more for real-time telephony application services, whereas WSML was designed more for data type services. The knowledge level between these two development mechanisms has little overlap. Therefore, a Web services version of Parlay, called *Parlay X*, was developed to allow the Internet developer simpler access to the operator's network and eliminates a Parlay learning curve. This has led to the development of Parlay X Gateways for deployment into an IMS network. Two types of implementations of the Parlay X Gateway, as shown in Figure 6.4, are the most likely. One is a stand-alone Parlay X

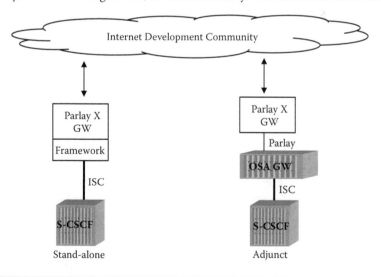

Figure 6.4 Parlay X Gateway implementation architectures.

Gateway for operators that only want to work with Parlay X application developers; here, the framework capabilities are onboard. The other option could be considered an adjunct model, with the Parlay X Gateway acting primarily as an interface translator and leaving most of the framework functions (e.g., authentication, security, SLA, etc.) to the OSA Gateway.

6.4 SCIM and Service Broker

The Service Capability Interaction Manager (SCIM) has been defined in 3GPP TS 23.002 as part of an AS (Figure 6.2). As discussed earlier in this chapter, very little is actually said about the function of the SCIM other than its role of an interaction manager. From Chapter 2, we examined the SCIM taking on the role of a switch's internal service interaction manager, which handled all the feature interaction issues for a switch's set of services.[5] We can extrapolate further from our architecture shown in Figure 6.2 the SCIM serving the role of extending the iFC process to manage the communication (i.e., interaction) between different applications in a blending role as well as in an interaction manager's role.

There is an effort under way within 3GPP to provide further definition of the SCIM. This is being done as part of a 3GPP Release 8 work effort entitled "Study on Architecture Impacts of Service Brokering." At the time of writing of this chapter, the effort (designated TRS23.810) is still only a study effort with the targeted output being a technical report (versus a technical specification) only [2]. *Service broker* as an alternative name for SCIM can lead to nomenclature confusion as currently they are being used interchangeably within the industry. The layman's definition is that the SCIM handles all interactions within an AS (as shown in Figure 6.2), and the service broker provides this same interaction functionality but more so as a stand-alone component (or network element). 3GPP is working to expand the scope of the service broker to work with other network technologies (e.g., CAMEL), but the definition is still in flux.

Given the late timing of both the SCIM and the service broker efforts, the industry has moved forward to bring definition to and develop product to the SCIM concept as a stand-alone product, with various vendors calling their product a SCIM and others calling it a service broker. For our purposes, we will use the term *service broker* when referring to the stand-alone network element and the term *SCIM* when referring to the blending function occurring within a single AS containing multiple applications. We will define here the scope of the service broker to support more than just the ISC (IMS) Interface and include the interface to other access networks, the most commonly discussed ones being CAMEL and Web Services 2.0 (although others such as AIN are being actively discussed by the vendor community).

In addition to handling feature interactions, it is a natural step for the service broker to play a blending role between multiple applications. Now, instead of just reacting to two different applications that may have a feature interaction issue, the service broker can be proactive in bringing disparate applications together for a richer user experience. We saw the beginnings of the blending service concept in Section 6.10 with the iFC. Where the iFC differs from the service broker is that the iFC can only work based on a static set of rules, namely, a list of ASs it should query in a fixed order with no variation based on an event or condition occurring in the network at that particular moment when an event is occurring with the user. On the other hand, the service broker is more dynamic in nature, which allows routing algorithms to change depending on dynamic conditions such as whether the user is currently on an active call or roaming internationally. In addition, it allows multiple ASs to be queried simultaneously, in contrast to the sequential nature of iFC.

So, we modify our example in Section 6.1 to add a "Find-Me" AS with Bob placing his office, laboratory, and mobile phone on the list for simultaneous alerting. We further expand our scenario with Bob forgetting to charge his mobile phone's battery, which causes his mobile phone to power off without his realization. In a typical simultaneous ringing scenario, under these conditions, the mobile phone would go immediately to voice mail with the voice mail platform responding with an answer message, which, in turn, would cause Bob's office and lab phone to never ring (or certainly no more than once with little opportunity to answer). Needless to say, Bob will be very upset to not receive any calls until he can go back home and charge his mobile phone.

This extended example involves both the iFC and service broker functions. The S-CSCF can send the initial query to the ringback tone application and then along to the service broker. The service broker can then send triggers to both the call logging and the simultaneous ringing ASs and place itself in the signaling path to receive responses from Bob's alerting list. If the service broker detects an answer message within a set period of time (e.g., less than one ring cycle) that would indicate our scenario, it can "hold" that message (what it does will vary by the service definition, but one scenario could be for the service broker to send a disconnect message back to the voice mail platform and let the other two phones ring). So, what we are seeing here is the service broker using its capability to take a proactive action in a service interaction scenario. In Section IV, we will look at some more examples of the Service Broker tying together multiple applications into a single service offering.

There is one last capability that the service broker could provide, one which is also getting some traction in the industry. We had examined earlier how 3GPP had tried to provide interworking with the legacy CAMEL network through the IM-SSF. On a separate note, there is a limitation within the CAMEL specifications in that it only allows a service to trigger to a single AS for a particular call state. If we couple these two points together along with the service broker's ability to facilitate interactions between different ASs, we gain a very powerful capability to now blend multiple CAMEL applications together without making costly network upgrades; furthermore, there is now the ability to blend CAMEL and SIP-based

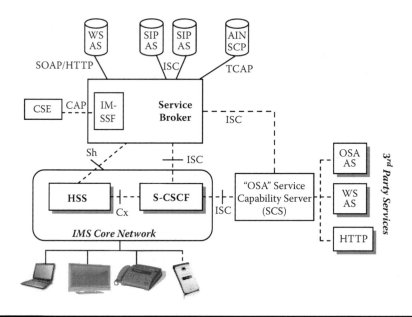

Figure 6.5 Service-broker-based converged IMS/CAMEL architecture.

applications together. So, even though the service broker is defined with a SIP (ISC) Interface, through proper separation of functions, one can add a CAMEL module (e.g., IM-SSF) to support CAMEL interworking while still keeping the same underlying resources to support a new technology. This philosophy can be extended to other existing networks such as third-party or Web services networks to allow blending of IMS applications with other technologies.[6] Figure 6.5 shows an example architecture for the service broker that could allow it to support CAMEL, SIP, and the blending of other technology services.

This figure varies from Figure 2.5 (Chapter 2) slightly because only the SCIM and not the service broker has been officially defined as part of the IMS architecture. The standards bodies are still wrestling with the definition of the service broker function even though the industry has already made its own decision.

Endnotes

1. To simplify our figure, we did not show the AS interface to the Subscriber Locator Function (SLF). In a multiple HSS network, the AS would go to the SLF first to find the correct HSS. In this case, instead of the Sh interface, it would be the Dh.
2. The IM-SSF is not a concept in a CDMA MMD network as CAMEL is defined only for GSM networks.
3. For more details than you could ever want, OSA is described in a multivolume set of standards in 3GPP TS 29.198 and TS 29.998.

4. More information can be found about the Parlay Group at www.parlay.org. At the time of this writing, it has been recently announced that the Parlay Group's work will be incorporated with OMA.

5. The classic example of a service interaction conflict is what happens when a subscriber has both call waiting and call forwarding assigned to their phone. The service interaction manager must resolve these conflicts.

6. It should be clearly noted that the effort to incorporate other IN technologies has not been standardized or even sanctioned yet in 3GPP (at the time of this writing). Early adoption of this approach will allow vendor and operator differentiation.

Further Reading

3GPP TS 23.002, "Network Architecture."

3GPP TR 23.810, "Study on Architecture Impacts of Service Brokering."

3GPP TS 23.008, "Organization of Subscriber Data."

3GPP TS 23.228, "IP Multimedia Subsystem (IMS) Stage 2."

3GPP TS23.278, "Customised Applications for Mobile network Enhanced Logic (CAMEL) Phase 4; Stage 2 IM-CN Interworking."

3GPP TS 29.198, "Open Service Access (OSA); Application Programming Interface (API)" Part 1–14.

3GPP TS 29.198, "Open Service Access (OSA); Application Programming Interface (API); Mapping for OSA" Part 1–8.

3GPP TS 29.328, "IP Multimedia (IM) Subsystem S_h Interface; Signalling Flows and Message Contents."

3GPP TS 29.329, "Sh Interface Based on Diameter—Protocol Details."

Chapter 7

Security

With constant stories appearing about computer hackers and other network breaches, security must be a top concern for any network operator as well as the end user. The migration to an Internet Protocol (IP)-based technology will expose the mobile operator and their users to security attacks similar to those attacks today associated with the wired Internet Service Provider (ISP). Third Generation Partnership Project (3GPP™) has defined two areas in which they address the security issue. In 3GPP TS 33.203, they lay down the security requirements between the IMS network and the end user (access security), and in 3GPP TS 33.210, they specify the security requirements in the core network domain (network security). Here, the core network domain would cover security between different IP Multimedia Subsystem (IMS) networks and also within a single IMS network.

As the IMS network "rides" on top of different packet networks (such as a General Packet Radio Service [GPRS] packet-switched network in the cellular environment), it offers an additional level of security protection against any breaches. For example, if for any reason an intruder breaches the security mechanisms of an operator's GPRS packet-switched network, this does not mean that the IMS network has been compromised. In fact, IMS provides a layer of separation from the underlying packet-switched network and provides its own unique layer of security protection separate from the access network transport layer.

7.1 IMS Security Architecture

In 3GPP TS 33.203, five separate security associations, each with its own need for security protection, have been defined. This model identifies where exchanges

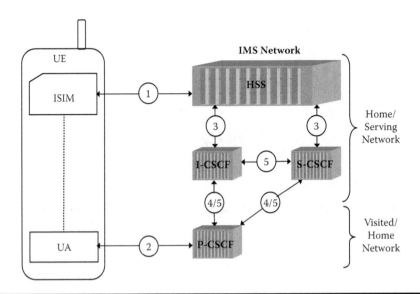

Figure 7.1   IMS security architecture.

in Session Initiation Protocol (SIP) messaging occurs and, thus, potential vulnerability points (Figure 7.1). The five security associations identified in TS 33.203 are as follows:

1. ISIM ↔ HSS: Required to provide mutual authentication. The home subscriber server (HSS) is provisioned with authentication keys and generates challenges. The IMS Subscriber Identify Module (ISIM) and the HSS contains one IP Multimedia Private Identity (IMPI) and at least one IP Multimedia Public Identity (IMPU) for each subscription. The shared secret key that is associated with the IMPI is contained in the ISIM and the HSS.
2. UA ↔ P-CSCF: Provides a secure link between the user equipment and the IMS network. The source of the data is verified to be the claimed subscription.
3. I/S-CSCF ↔ HSS: Provides a network security association for the transfer of information between the Call Session Control Function (CSCF) and the HSS.
4. P-CSCF ↔ I/S CSCF: This network security association is applicable only when the P-CSCF is in another carrier's network (i.e., visited network). If the P-CSCF is part of the home network, then security association #5 would apply.
5. CSCF ↔ CSCF: Provides a network security association between SIP capable nodes that are within the same network (i.e., home network).

7.2 Access Security

Access security covers two primary aspects: how the SIP signaling is secured between the IMS network and the end user, and how each end (network and end user) can be assured that the entity with which communication is desired to be established is actually the entity that it claims to be.

7.2.1 Universal Integrated Circuit Card (UICC)

Security functions within the end user's device are based on the information contained on the smart card (or commonly referred to as the SIM card). Use of the term *SIM card* is a carryover from the Second Generation (2G) world and is often used interchangeably with the more accurate 3GPP terminology—UICC (Universal Integrated Circuit Card).

The UICC is a multiapplication repository for 2G and 3G applications as well as specific vertical applications (Figure 7.2). It can contain a SIM application for backward compatibility with 2G devices that only support the SIM. It can contain a USIM application that is used by 3G devices and newer 2G devices. It can also contain an IMS Subscriber Identity Module (ISIM) application that holds the set of IMS security data and functions on the UICC for establishing a session with the IMS network (such as authentication keys, subscription, and ID information). Every ISIM contains a shared secret key that is also stored in its home HSS.

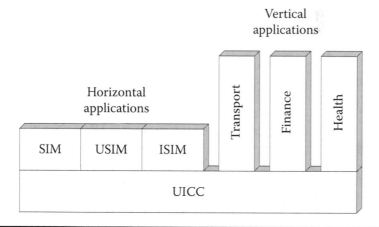

Figure 7.2   UICC architecture.

There may be more than one USIM or ISIM application on the UICC based on different network operator parameters; however, only one SIM application can reside on a UICC.

Although the use of the ISIM application on the UICC is the recommended method for access to the IMS, it is not a requirement. One alternative method includes using the security mechanisms associated with the USIM application. Another alternative method is the use of the Hypertext Transfer Protocol (HTTP) digest procedures as specified in RFC 2617. The early attraction of these alternatives was a time-to-market advantage, as vendors had not implemented the ISIM approach or an operator did not have ISIM-capable UICCs in their inventory. Although both alternative methods provide a level of security for accessing the IMS, they fail to provide a layer of security equal to that of the ISIM. or they greatly increased the complexities on the operator's part in managing the authentication data.

7.2.2 Registration of an IMS Subscriber

When a user powers on his or her device or moves between cell sites, there is a mutual need for the network operator to know that the device has a legitimate subscription and for the user to know that the network he or she is connecting to is a legitimate network site.[1] The IMS registration establishes a set of challenges and known information exchanges in order to authenticate both ends of the session establishment. The scheme for exchanging authentication information in the IMS is known as Authentication and Key Agreement (AKA) The different parameters that make up the AKA scheme are shown in Table 7.1. The IMPI (private identity) is the means used to uniquely identify the subscriber to be authenticated. In turn, this identifies which shared secret key (K) must be validated. In addition to the shared secret key, there is also a sequence number (SQN) that prevents replay attacks (i.e., capturing and resending of a valid authentication vector) by a false IMS network. To further secure the authentication process, the secret key is never sent between the ISIM and the HSS. Instead, different parameters are generated based on known algorithms and the secret key, which are then exchanged to authenticate both the user and the network. The authentication vector is temporary data that enables the IMS network to perform the AKA authentication process with a user and consists of the random challenge (RAND), response/expected response (RES/XRES), cipher key (CK), integrity key (IK), and authentication network (AUTN) parameters. The algorithms that determine the values of these parameters are beyond the scope of this text. What the reader should understand is there is a unique secret key that is known only to the ISIM and the home network (HSS) that is required for the authentication process. For readers desiring to understand the complexities of the AKA process, 3GPP TS 33.102 provides a full description.

The call flow for the IMS registration process is shown in Figure 7.3. For the purposes of simplicity, it is assumed that the user equipment (UE) has completed the prerequisite registration signaling for a GPRS attach and has obtained a Packet

Table 7.1 Parameters used in AKA calculations

AKA parameter	Length	Description
K	128 bits	Secret key: authentication key shared, exchanged between the HSS and the ISIM
SQN	48 bits	Sequence number, tracking the sequence of the authentication procedures to prevent replay attacks
RAND	128 bits	Random authentication challenge generated by the network
RES/XRES	4–16 octets	Authentication response generated by the ISIM
CK	128 bits	Cipher key generated during the authentication process
IK	128 bits	Integrity key generated during the authentication process
AUTN	128 bits	Authentication (Network) token
AUTS	112 bits	Synchronization token generated by the ISIM upon detecting a synchronization failure

Data Protocol (PDP) context. After this occurs, the UE sends a SIP REGISTER message to the P-CSCF (1). It contains the authentication information, including the private user identity (e.g., user1_private@att.com). Because the UE has not been authenticated at this point, the P-CSCF can only act as an outbound SIP proxy. The P-CSCF in turn forwards the REGISTER message to the I-CSCF (2). As we are assuming this is a new registration, the I-CSCF must query the HSS to determine the information necessary with which to assign the appropriate S-CSCF for the UE (3). This is accomplished via the exchange of the DIAMETER messages: User-Authorization-Request (UAR) and User-Authorization-Answer (UAA). With the S-CSCF assigned, the I-CSCF directs the REGISTER message to the S-CSCF (4). The S-CSCF must then obtain a number[2] of authentication vectors for the UE from the HSS (5) via the exchange of the DIAMETER messages Multimedia-Auth-Request (MAR) and Multimedia-Auth-Answer (MAA). Each vector received contains the following five parameters listed in Table 7.1: RAND, XRES, CK, IK, and AUTN. The AUTN is created by the HSS using the shared secret key (K) and the sequence number (SQN), which is kept synchronized between the ISIM and the HSS. The S-CSCF selects the first authentication vector received to send an authentication challenge to the UE via the SIP 401 Unauthorized message (6). When the P-CSCF receives the 401 Unauthorized message (7), it removes and stores the keys (IK, CK), and forwards the rest of the message to the UE (8).

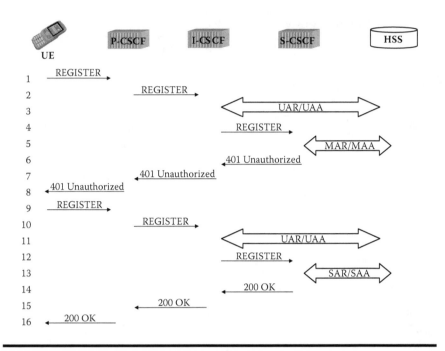

Figure 7.3   Registration and authentication call flow.

When the 401 Unauthorized message is received at the UE, the UE performs a check on the AUTN according to 3GPP TS 33.102 to authenticate the network. Once authenticated, the UE calculates a response value (RES) based on its secret key and the received RAND. The RES is returned to the S-CSCF in a second REGISTER message (9, 10, 12). The I-CSCF must query the HSS again to determine the S-CSCF (11). At the S-CSCF, the active XRES for this UE is retrieved and checked against the RES valued for a match. If the check succeeds, then the UE has been authenticated by the network. Because we will assume here that the IMPU of the UE has not been registered with the HSS, the S-CSCF must register it (13) using the DIAMETER messages Server-Assignment-Request (SAR) and the corresponding Server-Assignment-Answer (SAA). Other IMPUs corresponding to the IMPI may be implicitly registered at this point as well. The S-CSCF confirms the registration process to the UE by returning the SIP 200 OK message (14, 15, 16).

7.3 Network Security

Network security in the IMS addresses two scenarios:

1. Security within the same domain
2. Security between different domains

Here, a security domain is any network that is managed or is under the control of a single administrative authority. A typical example of a security domain would be a single network operator within a single country. Thus, an example of security between network domains would be a data exchange between two different network operators, as might occur in a roaming scenario. At the border of each security domain are security gateways (SEGs) that are used to secure the IP-based traffic. Because all IP traffic between security domains must go through the SEG, this will result in all traffic passing through at least two SEGs. Figure 7.4 shows examples of the security domains and the SEG.

3GPP TS 33.210 has defined interfaces for each of these two domains, as shown in Figure 7.4:

Za—Addresses all IP traffic between different domains. For this interface, the implementation of authentication and integrity protection is mandatory, and encryption of all IP traffic is highly recommended.
Zb—Addresses all IP traffic between the SEG and a network entity (NE), such as the CSCF, and also between NEs within a single security domain.

Because the Zb interface is completely under the control of a single operator, it is defined as an optional interface, leaving the decision up to individual operators to choose to implement their own best-practices policies instead of those defined by 3GPP. Conversely, the Za is a mandatory interface because it goes between different operators' security domains.

The traffic over the Za interface (and Zb if implemented by the network operator) is protected through the use of the IPSec ESP (Encapsulating Security Payload)

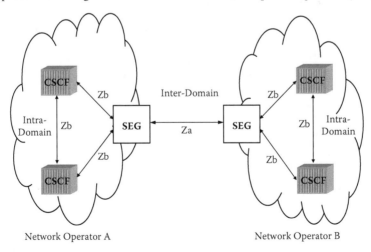

Figure 7.4 Network security architecture.

as described in RFC 2406 operating in tunnel mode. The negotiation of the IPSec Security Association between the SEGs is accomplished by using the Internet Key Exchange (IKE) as specified in RFC 2409.[3]

Endnotes

1. Bogus cell sites are a known security threat area where information about the user is fraudulently obtained.
2. 3GPP TS 33.203 states: "Upon receiving the SIP REGISTER the S-CSCF shall use an Authentication Vector (AV) for authenticating and agreeing a key with the user. If the S-CSCF has no valid AV then the S-CSCF shall send a request for AV(s) to the HSS in (UAR) together with the number m of AVs wanted where *m* is at least one." The value of m is defined by the operator based on their best-practices procedures every time. More than one vector is downloaded at once to avoid loading down the HSS with requests everything the S-CSCF needs to authenticate the user.
3. It should be noted that both RFC 2406 and RFC 2409 have been made obsolete by RFC 4303 and RFC 4306, respectively; however, 3GPP procedures state that RFC references cannot assume to be replaced by an RFC that has superseded it. This is well noted in our reference earlier; RFC 4306 is emphatic that IKE v2 (which it describes) is not backward compatible with IKE v1 as described in RFC 2409. In this example, if an IMS equipment vendor assumed the superseded RFC and developed to IKE v2, then its product would be incompatible with IMS equipment deployed by another IMS operator whose vendor developed to the stated RFC reference (and implemented IKE v1).

Further Reading

3GPP TS 33.102, "3rd Generation Partnership Project; Technical Specification Group Services and System Aspects; 3G Security; Security Architecture."

3GPP TS 33.203, "3rd Generation Partnership Project; Technical Specification Group Services and System Aspects; 3G Security; Access Security for IP-Based Services."

3GPP TS 33.210, "3rd Generation Partnership Project; Technical Specification Group Service and System Aspects; 3G Security; Network Domain Security; IP Network Layer Security."

RFC 2406, "IP Encapsulating Security Payload (ESP)," Kent, S. and R. Atkinson, November 1998.

RFC 2409, "The Internet Key Exchange (IKE)," Harkins, D. and D. Carrel, November 1998.

RFC 2617, "HTTP Authentication: Basic and Digest Access Authentication," Franks, J., Hallam-Baker, P., Hostetler, J., Lawrence, S., Leach, P., Luotonen, A., and L. Stewart, June 1999.

Chapter 8

Charging

Charging (and billing) is probably the most important function of a service provider as it supports the company's entire economic structure; yet, it remains one of the least glamorous parts of the network. In this chapter, we hope to shed some light on the "glamorous" aspects of how new classes of services can be enabled through a flexible charging mechanism provided through the 3GPP™ Internet Protocol (IP) Multimedia Subsystem (IMS) charging architecture. We will also examine the components of the IMS charging architecture.

8.1 Charging Capabilities

In Chapter 1, we identified *flexible charging* as one of the value propositions that come with IMS. In this chapter, we will list many of the principles underlying the design for charging and billing requirements. Behind the design for these requirements is the thinking that new services could be offered solely on the basis of a new or different way to charge. Part of these different views showed themselves in the simple definition change of the common TLA (three-letter acronym) CDR, which was changed early in the Third-Generation (3G) requirements phase from call detail record (reflective of a circuit-switched model) to charging data record (more reflective of an IP or data model). When we look at charging capabilities, we will examine the different ways an operator can charge for a particular service. In pre-IMS networks, typically, only a single charging method was used during a particular call or data session. With IMS, more than one charging method may be used because we are entering the world of multimedia multiple-session services. So, with these thoughts as background, let us look at some of the different ways in which charging can be implemented.

Pre-IMS networks (and those adopted by IMS networks) have all implemented some combination of the following models to assign charging responsibilities:

■ Calling Party Pays
  – For their leg of the call to the called party's home switch.
  – For both their leg and the called party's leg of the call.
■ Called Party Pays
  – For the call leg from their home switch to their current location.
  – For all legs of the call (as in the case of an 800/freephone number).
■ Zero-Rated Call
  – The CDR is assigned a zero charge rate on the basis of some marketing or other promotion package.

To collect their revenue from the customer, operators have implemented one of the following two models:

■ Prepay service (or *online* charging): Customers have purchased ahead of time a certain amount of usage that is debited for each use prior to network resources being used. Here, charging must be done on a real-time basis as the need occurs.
■ Postpay service (or *offline* charging): Customers have established a credit account with their service provider to be billed (typically) on a monthly basis for charges incurred. Here, charging can be done on a non-real-time basis.

The richness of IMS applications can begin to tax these models with their ability to offer multimedia services. One obvious example is a service that invokes multiple mediums such as a video share call (voice call with video streaming between end-points). The customer of this service may wish to only use the video streaming service portion sparingly because of the associated premium charge. Thus, the customer may wish to set up a separate account to limit his monthly spending on the video share service through a prepay account.[1] However, as he or she has a generous bucket of minutes for voice calls, the customer may want to continue those charges with his or her normal postpay service. What this example shows is that there is now an inherent capability to be able to charge separately for the different types of mediums employed during the session (remember, the concept of a call has been replaced by that of a session to reinforce the IP application concept) as well as by the actual service invoked (e.g., our video stream service example may be charged differently than from a data file transfer even though both services are essentially just passing data packets). The service provider can also choose to charge on a per-incident basis (event-based charging) rather than on monitored-time basis (session-based charging).

Further expanding on this aspect, event-based charging can take on many forms depending on the commercial offering of the service provider. The obvious example would be the delivery of a short message (Short Message Service [SMS]) or a multimedia message (Multimedia Messaging Service [MMS]), in which the operator charges for an event, that is, the delivery of a set message having a maximum fixed length. It can also be applied to "open-ended events" such as the previously mentioned video-sharing example. Here, a service provider could charge a set rate for associating the video stream with a call independent of the length of the call. Whether this is actually offered by a service operator is dependent on the operator; however, the point is that operators are allowed to be creative and flexible in how they charge for their services.

Other charging permutations identified in 3GPP TS 22.115 include the following:

■ Charging for different levels of quality of service (QoS) during a session on a per-medium or per-service basis.
■ Charging separately for individual legs of a session: This also applies when multiple mediums are invoked (e.g., voice call with video streaming).
■ User can be charged on the basis of the current access technology that the service is invoked with (e.g., Wi-Fi access may be zero rated, but a Second Generation [2G] access call is charged at standard rates).
■ Conversely, the user can be charged on the basis of the service invoked, independent of the access technology (e.g., an SMS message delivery is charged 20¢ regardless of whether it is delivered over a Wi-Fi connection or over a Universal Mobile Telecommunications System [UMTS™] network).
■ Charging while roaming can be done with the same method as for the home network (e.g., duration charging for video sharing in the home market can also be applied while the user is roaming).
■ Charging can be applied on the basis of location, presence, or whether the service is pushed to the user.

Other considerations and models in charging that are needed in a converged IP environment are the following:

■ Wholesale charging
  – Wholesale charging takes place between the originating and receiving operators and service providers.
  – Typically, operators and service providers identify and record the charging event, and then look up the interworking business agreement to determine the rate for each charging event.
■ Group versus individual messages: Charging rates can be different for one-to-one and one-to-many (multiple recipients) messages.
■ New charging models such as Pay for Subscribe, Privilege to be Notified, and Pay for Notification (e.g., when Presence status changes).
■ Participant Party Paid (PPP): This is identified as an alternative means for services such as Push-to-Talk over Cellular (PoC) where charging in addition to

Calling Party Paid (CPP) is applied. Third-Party Paid is an example in which a third party, such as an advertiser typically in location concierge services, may be charged for the event.

As we look at charging in more detail, we should also add that IMS will extend current business relationships. Figure 8.1 reflects the view of the business relationships between the mobile network operator and other business partners as laid out in 3GPP TS 22.115. Wireless local area network (WLAN) operators are the obvious new business partners. Third-party vendor relationships are not new as those between business partners and operators; however, in the past, they have been "one-off" relationships requiring specially dedicated connections. IMS has defined an interconnection reference point to allow third parties to access the operator's network resources as well as to provide services to the end user. This will eliminate the need to have specially dedicated resources per third party or per service as well as the ability to allow the operator to provide third-party service level agreements (SLAs) (see Chapter 6, Section 6.3 for additional details).

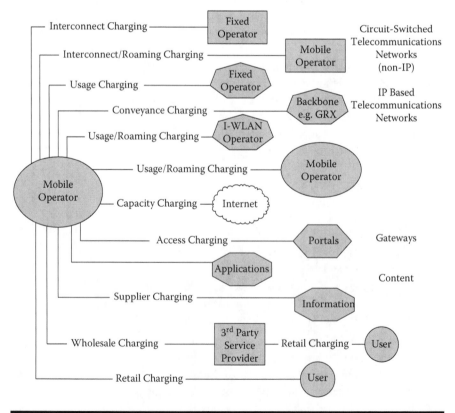

Figure 8.1   Charging entities relationships.

Table 8.1   Third-party charging items

Item	Charging method description
Content accessed	The third party charges the end user for content accessed or downloaded and uses the network operator as its billing agent.
Access to site	The third party is charged by the network operator for each invocation by the end user.
Location information	The network operator charges the third-party service provider for location information related to the end user.
Presence information	The network operator charges the third party for presence information about the end user.
Pushed information	The network operator charges the third party for each message (e.g., text, video, etc.) pushed to the end user.

Third-party applications will be instrumental in bringing the service richness of the Internet to the IMS user. To achieve this goal, both the network operator and the third-party service provider must be able to accurately charge each other for the use of their respective resources. Routing of these resources through a mediation gateway such as a Parlay or Parlay X Gateway will permit service control, accurate metering, and a guaranteed service level (e.g., minimum support for X transactions/s). Previously dedicated connections (such as a Short Message Peer-to-peer Protocol [SMPP] link to send SMS-based information services) did not have this capability, which prevented SLAs from being offered. Some items identified from 3GPP TS 22.115 that can be a basis for charging are listed in Table 8.1. Certainly, operators are free to (and do) specify other parameters.

Having accurate CDRs allows for accurate settlement of accounts with the third-party service provider and can aid the operator in accurately settling disputes between the end user and the service provider.

8.2 Charging Architecture

The 3GPP Charging Architecture is defined in a series of specifications, each designated to cover a particular domain, enabler, or format. Figure 8.2 shows the different 3GPP charging specifications. Although components of many of these specifications will have synergies with an IMS deployment (e.g., a PoC service blended with another IMS service), our focus will be on the charging architecture for an IMS network and the two types of charging available, namely, offline and online.

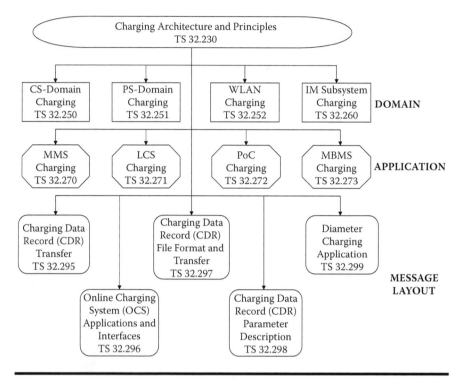

Figure 8.2   3GPP charging standards.

8.2.1 Offline Charging

The 3GPP Charging Architecture is defined for Release 5[2] in 3GPP TS 32.200 and for Release 6 (and beyond) in 3GPP TS 32.240. An architecture change and renaming occurred between Release 5 and Release 6; we will address this as the renaming can cause some confusion. The Release 5 offline charging architecture is shown in Figure 8.3. Each of the IMS elements shown has a direct connection over the Rf interface to the Charging Collection Function (CCF). The CCF acts as the receiver of all charging information from the IMS elements and then constructs and formats the information into a CDR, which it then passes to the billing system (BS) over the Bi interface. The BS will end up creating the final billing record (or CDR) on the basis of the charging information received from all sources (CCF and Charging Gateway Function [CGF]).

CGF provides a similar function as the CCF, namely, a single reference for interconnection, which in this case would be to the Gateway GPRS (General Packet Radio Service) Support Node/Serving GPRS Support Node (GGSN/SGSN), and performing the preprocessing necessary to forward the CDR to the BS as well as any needed buffering in case the BS is overloaded or experiences a temporary outage.

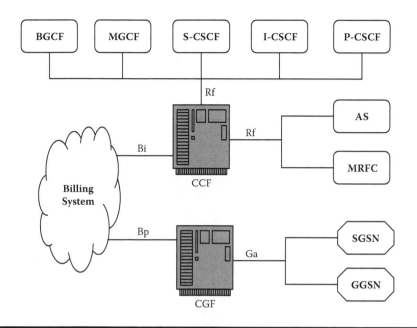

Figure 8.3 Release 5 offline charging architecture.

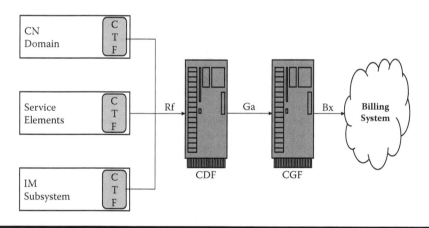

Figure 8.4 Release 6 (and beyond) offline charging architecture.

Release 6 (and beyond) defines a common type of egress point across all networks and service elements called the *Charging Trigger Function* (*CTF*), which connects to the Charging Data Function (CDF) over the Rf interface (see Figure 8.4). Within the IMS, all the control function elements (P/C/I-CSCF, Breakout Gateway Control Function [BGCF], Media Gateway Control Function [MGCF], and Media Resource Function Controller [MRFC]) and the Session Initiation Protocol Application Server (SIP AS) implement the Rf interface. Examples of elements in other domains supporting the Rf interface would include mobile switching center (MSC), service control point (SCP), GGSN, SGSN, or WLAN.

The CTF will generate charging events on the basis of resources invoked by the user after collecting that information from the network or service element. CTF is an integrated component for network elements providing offline (and online) charging. It has two primary functions:

- Accounting metrics collection: Monitoring of events to provide metrics that identify the user and his or her usage of network resources
- Accounting data forwarding: Assembles chargeable events from the collected metrics and forwards them to the CDF over the Rf interface

CDF performs many of the functions that were characteristic of the CCF, namely:

- Single (logical) element to receive charging information (from all domains) for all network elements (albeit through the CTF) over the Rf interface
- Construction and formatting of the preprocessed CDR on the basis of charging information received

Now, instead of going directly to the BS like the CCF (in Release 5), the CDF forwards the CDRs generated by it to the Charging Gateway Function (CGF) over the Ga interface. The CGF acts as a gateway between the IMS network and the BS, aggregating input from one or more CDFs. It performs any necessary preprocessing on the CDR, including validating and performing error handling, filtering, buffering, and file transfer to the BS.

The Release 5 specification clearly states that its thinking on the CCF was "for further study." We see the maturing of that thinking in the Release 6 (and beyond) specifications with the additional architectural definitions of CTF and CDF.

8.2.2 Online Charging

The connection interfaces between the IMS and the prepaid charging system is limited to just three network elements: AS, MRFC, and the S-CSCF. As shown in Figure 8.5, the AS and MRFC connect to the Online Charging System (OCS) over the Ro interface. The S-CSCF uses the IMS Service Control (ISC) Interface and connects through an interworking gateway (IWS-GWF) to present the Ro interface to the OCS.

The CTF is part of the online charging architecture, similar to the CTF, which is part of the offline charging architecture. However, instead of connectivity to the CDF, the CTF connects to the Online Charging Function (OCF) via the Ro interface, as shown in Figure 8.6. Also, similar to the offline charging architecture, an architecture name change and definition enhancement occurred while going from the Release 5 specification to the Release 6 specification for online charging. The

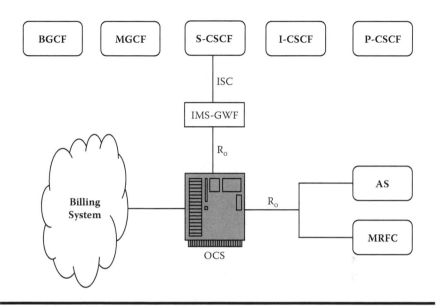

Figure 8.5   Online charging architecture for IMS.

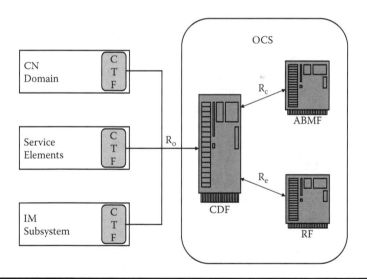

Figure 8.6   Release 6 (and beyond) online charging architecture.

OCF shown in Figure 8.6 was originally designated *Event Charging Function (ECF)* and is still referred to as ECF in certain 3GPP and Internet Engineering Task Force (IETF) specifications.[3]

Also, two additional parts of OCS shown in Figure 8.6 are the Account Balance Management Function (ABMF) and the Rating Function (RF). The ABMF is the location at which the subscriber's account balance can be found. The RF determines the charge for using a particular network resource. Together, the ABMF and RF along with the relevant policy information will inform the service control (the Policy and Charging Rules Function [PCRF], and Policy and Charging Execution Function [PCEF] discussed in Section 8.2.3) mechanism if the subscriber has enough credit balance to start or maintain a service or if the service needs to be denied.

8.2.3 Charging and Policy

Charging and policy are distinct, yet related. Charging involves the application of rules for event- and session-based transactions to apply business behavior that is primarily monetary in nature, such as the determination and application of a rate, notification of an advice of charge, decrementing a balance, or decreasing a counter for promotional minutes or free uses of a service. Policy allows for access control (e.g., Digital Rights Management), resource control (e.g., bandwidth, QoS, etc.), and routing control (e.g., content filtering).

Policy and Charging Rules Function (*PCRF*) is defined by 3GPP (Figure 8.7) as a repository of basic charging rules and policy rules in 3GPP Release 7 and onward. A common framework supporting complex transactional behavior such as charging is combined with policy request or response mechanisms for policy decision enforcements, and gives a new meaning to the policy-based charging paradigm.

The framework takes into account in-session events and subscriber profiling before authorizing or charging. The control component of the framework is called the *Policy and Charging Execution Function* (*PCEF*).

8.2.3.1 QoS-Based Charging

The PCRF and PCEF capabilities described in the previous section provide comprehensive and flexible policy execution that includes precise QoS control. The following are the QoS capabilities:

■ Authorize QoS resources at Application Function (AF) session establishment
■ Authorize QoS resources at AF session modification
■ Provision authorized QoS to the PCEF for execution

When a session is established for a service, the PCRF determines, on the basis of information on the service from the AF and subscriber details from the Subscriber Profile Repository (SPR), the appropriate QoS to be set up for the session. The PCRF then provisions the appropriate QoS to the PCEF, which in turn sends the required details of the provisioned QoS to the charging function.

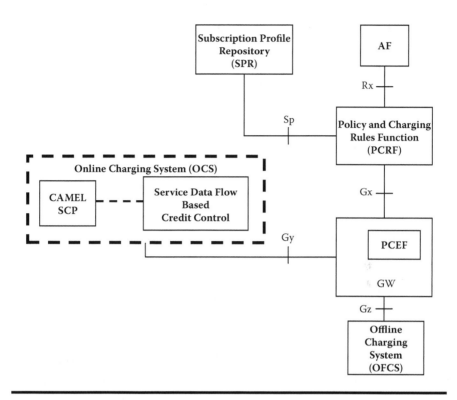

Figure 8.7   Policy and charging combined—as defined in 3GPP 23.203.

8.3 P-Headers for Charging

RFC 3455 defined a set of private header (P-Headers) extensions for use in 3GPP networks. Of immediate interest among these extensions is the P-Charging-Vector header, which is defined to allow operators to charge for access, resources, and services invoked by a user. The P-Charging-Vector is used to correlate charging records as a session, and its resources are invoked between different network elements. Also of interest is the P-Charging-Function-Address, which specifies where to send the charging information and infers whether to bill the session as an offline (postpay) or online (prepay) session.

IMS elements will generate charging correlation information and place it in the P-Charging-Vector, which consists of the following three parameters:

1. IMS Charging Identifier (ICID): A globally unique identifier of a session or transaction. It is used to correlate related charging data generated in different IMS elements.
2. Access Network Charging Identifier: Identifies the bearer used to access the IMS network. For example, the GGSN address and Packet Data Protocol

(PDP) context identifier for packet switch access or the Multimedia Charging Identifier for a fixed broadband access.
3. Inter-Operator Identifier (IOI): A globally unique identifier that identifies both the originating and terminating networks associated with a session or transaction.

These three parameters will uniquely allow an operator to track how a customer uses the different functions of an IMS network so that the user can be properly and accurately charged.

The P-Charging-Function-Address header contains the address of the primary CDF (or the OCF or both) to which the CTF should send the charging information for preprocessing. A secondary address may also be included if the operator desires to have a further level of granularity or to allow for identification of instances of CTFs in case of multi-instance deployments.

Knowledge of which charging element (CDF or OCF)[4] to send the charging information to is contained in the user's subscription profile in the Home Subscriber Server (HSS). Upon registration, this information is passed to the S-CSCF, which in turn passes the information to the different IMS elements within the home network. Note that based on the robustness of the operator's network, the S-CSCF can send out both the CDF and OCF addresses to different elements to support our example in Chapter 8, Section 8.1 about allowing the user to have both postpay and prepay services.

Endnotes

1. Flexibility is required of the 3GPP operator to allow the customer to have multiple prepay accounts for different kinds of services (e.g., Web access, mobile commerce [m-commerce], messaging, etc.).
2. The first true IMS deployments will follow the Release 5 specification.
3. For example, specifications referring to the P-Charging headers described in Chapter 8, Section 8.3, such as RFC 3455 and 3GPP TS 24-229 (Release 7) refer to the CCF and ECF instead of the CDF and OCF, respectively.
4. Note that other 3GPP specifications (e.g., TS 24-229) still use the Release 5 CCF and ECF terminology in referring to the IMS Charging Architecture.

Further Reading

3GPP TS 22.115, "Service Aspects; Charging and Billing."
3GPP TS 24.229, "IP Multimedia Call Control Protocol Based on Session Initiation Protocol (SIP) and Session Description Protocol (SDP); Stage 3."
3GPP TS 32.225, "Telecommunications Management; Charging Management; Charging Data Description for the IP Multimedia Subsystem (IMS) (Release 5)."

3GPP TS 32.240, "Telecommunications Management; Charging Management; Charging Architecture and Principles."

3GPP TS 32.260, "Telecommunications Management; Charging Management; IP Multimedia Subsystem (IMS) Charging."

3GPP TS 32.295, "Telecommunications Management; Charging Management; Charging Data Record (CDR) Transfer."

3GPP TS 32.296, "Telecommunications Management; Charging Management; Online Charging System (OCS) Applications and Interfaces."

3GPP TS 32.298, "Telecommunications Management; Charging Management; Charging Data Record (CDR) Parameter Description."

3GPP TS 23.203, "Technical Specification Group Services and System Aspects; Policy and Charging Control Architecture."

RFC 3455, "Private Header (P-Header) Extensions to the Session Initiation Protocol (SIP) for the 3rd-Generation Partnership Project (3GPP)," Garcia-Martin, M., Henrikson, E., and D. Mills, January 2003.

Chapter 9

User Equipment and IMS

In many ways, the most important aspect of an Internet Protocol (IP) Multimedia Subsystem (IMS) application is placed in users' hands. The mobile device, also known as user equipment (UE) in Third Generation Partnership Project (3GPP™) specifications, and the IMS applications it contains are the parts of an IMS-based service that end users will experience. It will be used to judge the overall success of any service offering. A well-designed, robust implementation in the UE is therefore critical to any IMS application deployment. The IMS architecture of the UE and some of the unique issues that come up when realizing an IMS application will be discussed in this chapter.

9.1 IMS Architecture in UE

9.1.1 Implementation of IMS Planes

When IMS is implemented in a UE, all three layers can be contained in the device. Each plane is provided by specific combinations of hardware and software:

- Bearer plane: Hardware and software that provides IP connectivity to one or more access networks.
- Control plane: This includes the Session Initiation Protocol (SIP) stack and the IMS Subscriber Identity Module (ISIM) application contained in the device's Universal Integrated Circuit Card (UICC).
- Application plane: The individual IMS enablers supported by the device.

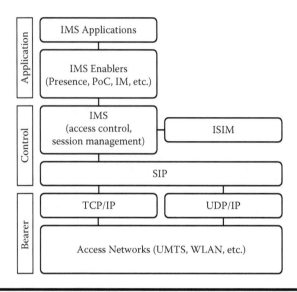

Figure 9.1   IMS layered structure in UE.

Figure 9.1 shows how the three layers of IMS map to hardware and software in the UE. Depending on the specific UE implementation, only the bearer and control planes may be provided, and the application plane (IMS Enablers and end-user applications) is developed separately.

Each layer in the UE will typically have standard or proprietary application programming interfaces (APIs) for applications or protocol stacks to use. The types of APIs and the layers they interface with will determine what functionality an IMS application or client has to implement by itself and what portions can be provided by the host UE.

9.1.2 Application Types

Now that we have an IMS-capable device, the type of application to be created needs consideration. Applications on mobile devices can usually be categorized into one of the following two types:

- Embedded application: Application is preinstalled on the mobile device.
- Downloaded application: Application is installed after the device has been manufactured.

Embedded applications may use native code and can be highly integrated with the host device's operating system. A downloaded application, however, generally must make use of published APIs such as those defined by Java ME (Micro Edition).

These standard APIs may not provide the level of integration that an embedded application would be capable of providing.

The advantage of a downloadable application comes later when client revisions are needed. Except in the case of open operating systems (OSs), it is generally easier to update downloadable applications than embedded ones. This may be important if bug fixes or new features need to be distributed to existing users.

9.1.3 High-Level versus Low-Level APIs

With either type of application, the APIs that are used to access core device services may vary. A client may implement the entire IMS application down to the SIP stack or rely on the host device to provide some or most of the lower-layer functionality. Low-level APIs may be present to allow access to an embedded SIP stack with the remaining pieces of the application or enabler being handled by the client itself. If IMS Enabler APIs are available in the device, a client need only implement the IMS application interfaces and rely on the device for the required enablers. Figure 9.2 illustrates the different IMS interfaces that may need to be included in an application, depending on the available APIs.

The higher the level of API that a client can use to implement an IMS application, the easier it will be to develop. Low-level API usage requires an in-depth knowledge of IMS and SIP, and means that all IMS security and other conventions must be implemented properly in the client. Care must be taken during development to avoid badly behaved clients, as the API itself will generally not limit application behavior. High-level APIs remove this concern from the application developer. However, a reliance on high-level APIs means that their implementation must be tightly controlled; otherwise, inconsistencies across devices and vendors can cause just as many application problems as in the case of low-level APIs.

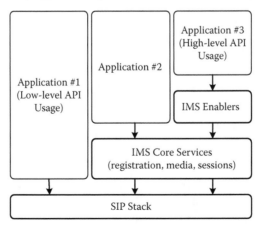

Figure 9.2   High- and low-level APIs for IMS.

9.1.4 IMS Frameworks and OMTP

The concept of high-level APIs has progressed to the definition of an IMS framework, a standard set of IMS capabilities that provide a robust application environment. An IMS framework may be native or Java based, and implements a standard set of IMS core and enabler APIs. The advantage of a Java or open OS-based framework is that it allows a single application to run on multiple devices. A proprietary OS implementation still has advantages such as better performance, but it requires each vendor to develop new applications for their specific platforms. By embedding an IMS framework into the device, all IMS clients now work like downloadable applications as they use standardized APIs. However, the APIs themselves are now highly capable and provide the integration and protocol support that would normally be limited to embedded applications.

The Open Mobile Terminal Platform (OMTP) forum is also defining a minimum set of requirements for IMS implementations on mobile devices. This work is not intended to define the specifics of the APIs themselves, but to standardize what functionalities an IMS framework should provide. By creating high-level definitions of the IMS core and services functionality that should be exposed to application developers, the goal of OMTP is to provide a consistent platform on which IMS applications can be developed. Included in these definitions are recommended profiles for different types of terminals, given that device memory, processor, and input/output (I/O) capabilities can vary greatly. As manufacturers start to adopt these recommendations, the ability to quickly implement new IMS services across a product portfolio will increase.

9.1.5 Java Environment

For clients implemented using the Java ME environment, there are different APIs defined that provide the range of low- to high-level capabilities described previously:

- Java Specification Request (JSR) 180: SIP API for J2ME
- JSR 281: IMS Services API
- JSR 325: IMS Communication Enablers

JSR 180 only provides a SIP stack for Java applications to access, and all IMS and application capabilities must be provided by the client itself. JSR 281 provides higher-level IMS capabilities but still requires an in-depth knowledge of IMS. JSR 325 provides access to a set of IMS enablers and provides the most complete environment to facilitate the development of IMS applications.

9.2 IMS Implementation Options

In addition to the basic choices of application type and APIs, there are several options available as to how an application, or even IMS and its security mechanisms, is implemented in a UE.

9.2.1 3GPP versus Internet Engineering Task Force (IETF) Mode

In the control plane, call flows for IMS sessions can generally follow one of two modes:

■ IETF mode: Session call flows as specified by IETF in the relevant RFCs
■ 3GPP mode: Session call flows as specified in 3GPP specifications

Although 3GPP IMS specifications were defined with the mobile environment in mind, they are generally more complicated due to additional signaling required to negotiate quality of service (QoS). Depending on the application, however, best-effort QoS may be sufficient to provide an acceptable user experience. In addition, most mobile networks have historically offered only best-effort QoS anyway. In such cases, IETF call flows will suffice and allow for a simpler client implementation.

9.2.2 Quality of Service (QoS)

If a specific level of QoS is needed for an IMS application, the selection of the QoS mechanism to use has a significant impact on client design and on the network. One option is to use the Session Description Protocol (SDP) described in Chapter 12 to negotiate QoS when operating with 3GPP-mode IMS signaling. For 3GPP networks, an alternative is to simply use the QoS options in the Radio Access Network (RAN) for primary and secondary Packet Data Protocol (PDP) contexts.[1] A third option is to use the Resource Reservation Protocol (RSVP) protocol. The available choices will depend on the access networks to be used and their QoS capabilities.

9.2.3 IMS Security and the ISIM

An authentication mechanism is essential to protect the IMS network from unauthorized users. Full 3GPP IMS security (3GPP TS 33.203) is available, but some service providers have chosen to use what 3GPP calls "early IMS security" (3GPP TS 33.978). Full security provides a much higher level of protection but at the expense of higher complexity. Internet Protocol Security (IPSec) can also be used in conjunction with full IMS security to add authentication and encryption at the IP layer.

The additional complexity and procedures needed for full security must be considered when choosing the type of IMS security to use. When full IMS security is implemented, ISIM authentication keys must be managed and loaded into the HSS so that the IMS Authentication and Key Agreement (AKA) authentication procedures can be completed. With the use of the ISIM to provide authentication credentials, the ISIM itself must now be provisioned with unique user information such as the Public User Identity (PUID). This may be a new process that needs to be implemented by the mobile network operator if the ISIM parameters are not known when the UICC is manufactured. Over-the-air (OTA) provisioning platforms and associated systems might be needed to provision a user's ISIM with the proper parameters.

Ideally, this provisioning process should be seamless to the user. It is an additional barrier to application usage if clients are not automatically ready to be used after provisioning. When provisioned over-the-air, there must be an additional OTA mechanism in place to automatically reload the updated ISIM information to the UE using an ISIM REFRESH command. IMS clients must be able to recognize this event so that users can immediately make use of the IMS capabilities in their device.

9.3 Access Network Issues

As mentioned in Chapter 1, IMS is defined as a bearer-agnostic method to deploy applications. However, the available networks can often determine how well an IMS application can run, if at all. The mobile client needs to be aware of these limitations and be designed to smoothly handle any network scenario it may face.

9.3.1 Bearer Suitability

Available bandwidth and latency will vary greatly among bearers (e.g., EDGE versus 802.11g), and applications may be sensitive to these differences up to the point that they are unusable on slower links. Even if an application can adjust its bandwidth usage to fit within what is available, it does not mean the result will be usable or acceptable to end users. Streaming of audio or video, for example, requires a minimum bandwidth and consistent latency that may not be available on some networks.

Clients will need to be aware of the capabilities of lower transport layers being accessed, and be able to make decisions based on the available networks. This may be seen as violating the Open Systems Interconnection (OSI) model in which each layer operates independently. However, in a multinetwork environment, each application may have different requirements that impact their ability to operate on one or more of the available networks. One does not want to run an application that is inappropriate for the network in use.

9.3.2 Handover between Bearers

For devices that support multiple access networks, applications may need to continue operation as the device moves from one network to another, such as between 3G and wireless local area network (WLAN). When a smooth transition is needed, one available option is to use Mobile IP so that a device's IP address remains constant as network changes occur. If longer breaks in service are acceptable when switching bearers, then other simpler methods can be used. The nature of the application itself and the desired user experience will determine how seamless this transition needs to be to the user.

Another issue to be considered is that combinational services require 3G coverage, assuming 2G DTM (Dual Transfer Mode) is not available. In such a scenario, combinational services will need to be gracefully terminated when handing over to other network types, including 2G networks of the same operator.

9.3.3 IMS Roaming

Applications running on the mobile network are usually envisioned to operate while roaming on other carriers' networks as well. However, a client needs to be able to handle scenarios in which no network is available to support the application. There may also be scenarios in which an operator may not want an application to roam based on specific use cases or roaming costs. It may be up to the client to either enforce these restrictions, or recognize when the service is unavailable and react in a predictable way for the user.

9.4 Mobile Device Considerations

Mobile devices can have unique issues that need to be considered during application and client design. From configuration of connection settings to the impact on device performance, there needs to be a plan to deal with these unique issues.

9.4.1 Client Provisioning

The handset must be configured with appropriate settings to ensure that the data traffic is routed appropriately for billing as well as service access. Configuration of the settings in the handset can be extremely complex as well. In a 3GPP network, there can be multiple Access Point Names (APNs) and PDP contexts available for data connections. Depending on the billing model for a service, it may be desired to have different kinds of data traffic be routed over different connections. Some examples of this might be the following:

- Single context: Route all services through one data connection.
- Multiple contexts based on service type: Services use different connections based on billing, or access or provisioning restrictions.
- Multiple contexts based on traffic type: Services route different types of traffic over different connections (e.g., signaling over one PDP context and media over another, etc.).

For the client, this means configuration of multiple connection profiles, which can introduce significant complexity. It is important that clients and mobile devices be customizable and be configurable OTA so that end users are not forced to make any manual changes. OMA Device Management (DM) and OMA Client Provisioning (CP) are two enablers that could be used for the OTA configuration, as average end users will be unable to manually configure their devices.

9.4.2 Connection Management

It is becoming common to have multiple data bearers available in one location, as well as devices that support multiple network types. The proliferation of 3G and personal, corporate, and public Wi-Fi means that applications and clients need to not only be aware of the available connections but also be able to choose the most appropriate one. There is not necessarily a "one-size-fits-all" network. Numerous criteria have to be combined to determine which bearer is the "best" among those currently available:

- Service access: Some services may only be accessible via one particular network type.
- Bandwidth and latency: What is the minimum network performance needed to support the service.
- Cost to user or network operator: Does one network result in higher costs to the user or operator versus another.
- Support of combinatorial services: Does the network support independent voice and data.
- QoS: Does the network support the QoS needs of the application, if any.
- Support for roaming: Is the service supported outside the home network.

When multiple access networks are supported simultaneously by the UE, the control plane needs additional functionality to manage the available connections and choose which one to use when setting up a session. This is illustrated in Figure 9.3.

In extreme cases, this leads to a diagram (Figure 9.4) where each application is mapped to a prioritized list of network types. This kind of logic further complicates the design, but it is necessary to keep the user from experiencing unexpected interruptions in service. Users also need to be sufficiently informed regarding the cost

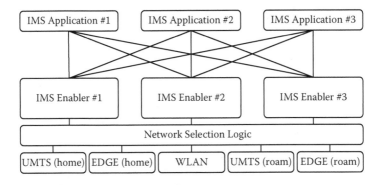

Figure 9.3   UE connection management.

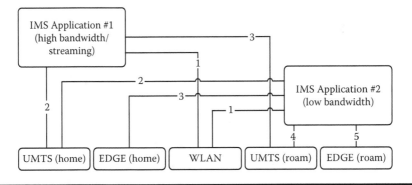

Figure 9.4   Application to network priority mapping.

of a particular service as the network in use changes. For example, WLAN service may be free, but 3G service is not.

9.4.3 Power Consumption

IMS applications can have a significant impact on power consumption and, therefore, the battery life of mobile devices. The long battery life typically available with mobiles is dependent on their turning off power-consuming components and processes for significant percentages of time. The radio transceiver is usually the highest power consumer in a mobile, so battery life is highly dependent on how often the radio components are in use.

IMS applications that require frequent data exchanges can prevent devices from using their sleep modes for maximum benefit. Presence is an example of a potentially "chatty" application, and the discussion in Chapter 16, Section 16.4, on traffic optimization describes the need to manage the amount of presence-related messaging. This optimization will also have a positive impact on the device's battery life.

Otherwise, if IMS traffic is not properly managed, a device may seldom be able to enter sleep mode, and battery life will be significantly degraded. Application design must trade off the user experience (e.g., how rapidly information is updated) versus the battery life of the device.

9.5 Client Implementation and Testing

One of the widely quoted promises of IMS is that new services can be introduced rapidly once an IMS infrastructure is in place. This statement is built on the assumption that a capable framework exists on which to introduce new services on end-user devices. Until the APIs and frameworks discussed in Section 9.1 are widely available, each new IMS service will take a significant amount of time to develop and test.

In all cases, however, it is important to have a standardized solution defined for an application. A complete and clearly defined set of IMS application requirements helps to minimize (but does not eliminate) potential issues with client interoperability. When interoperability between service providers is needed, standardization becomes even more critical. It is rare to find an application that was never envisioned to eventually be cross-carrier and roaming enabled.

9.5.1 Single Vendor or Multiple Vendors

In general, there are two different ways in which clients can be developed and introduced:

- Single vendor: A single vendor develops the client, which is then integrated or downloaded onto each device.
- Multiple vendors: Each device vendor develops or purchases their client, which is designed according to a common specification.

In the single-vendor case, a standard client and framework are defined that will be integrated into all devices. With multiple vendors, there is increased reliance on the application specification that all vendors must adhere to, but development is not tied to a single source. Each of these approaches carries benefits and risks in terms of cost, development time line, and integration effort that have to be evaluated on a case-by-case basis.

However, the amount of testing needed can be significantly different between the single and multiple vendor approaches to clients and frameworks. Interoperability testing of IMS clients, especially those supporting multiple services, is a complex undertaking. The greater the number of independently developed clients that exist for a single application or service, the more testing is required to verify interoperability.

9.5.2 *Interoperability Testing*

The testing process and the time line needed to launch an IMS application should not be underestimated. With any new service, unexpected bugs will be found in testing even with rigorous requirement process controls in place. Because many IMS standards are relatively new with few commercial deployments to date, significant testing is still needed to validate clients. Although standardized conformance and protocol tests have been developed for IMS enablers including those mentioned in Part III of this book, unique or peer-to-peer applications deployed by a service provider will require extensive custom testing regimens. In all cases, end-to-end interoperability testing is needed as well to ensure that clients will work properly when deployed.

End-to-end interoperability is the true test of an IMS application. It fills the gap left by protocol and conformance testing, and covers scenarios that can occur when different clients attempt to operate on actual access networks. These include operation in different network environments (e.g., networks with varying coverage and congestion), different access network types and manufacturers, and even across different service providers if applicable.

The variety of clients available for a single application often accounts for the biggest interoperability issues as opposed to core application design problems. Consider the case of multiple vendors implementing the same application independently. As each vendor introduces its client, ideally each new client needs to be tested against each existing one. This causes an increase in testing time and effort as the number of unique clients grows. At some point, this interoperability testing may become too large an effort to manage.

In theory, interoperability could be tested against only a representative set of clients to reduce the effort involved. However, this must be done carefully when peer-to-peer services are involved. A client-server application is a much simpler case as the network component is the same for all clients, and application servers (ASs) may normalize the message format and content forwarded to the receiving party. For peer-to-peer services, each client must be able to deal with the inconsistencies of every other client. From the order of SDP parameters to the usage of optional SIP headers, there can be numerous differences in SIP signaling that can create issues between clients. A detailed knowledge of the IMS application itself and the uniqueness of every available client must be considered when deciding the scope of interoperability testing required.

9.5.3 *Error Handling*

A client designed solely around expected network and application scenarios will likely have issues when deployed. Error handling is a critical part of the client design process and specification development. In addition to cases of varying signal strength impacting throughput, there can be unexpected disconnections of voice or data connections, lost packets, or errors generated due to maintenance or outages of equipment. Although in many cases service interruption cannot be avoided, a user should not be placed in a situation in which his or her application permanently stops working. Client error handling should be designed to allow automatic recovery whenever possible, so that the user does not have to perform manual steps (e.g., power cycling) to get an application working again.

There are four general areas that errors will fall into, each posing different challenges:

■ Application errors
■ IMS errors
■ Transport errors
■ Provisioning errors

Application error handling is usually well thought out as part of the application design. It is the lower-level IMS and transport issues that can have significant impacts if not considered in the design process.

IMS and transport layer errors can occur in a variety of ways. Such errors may involve longer-term outages in the IMS core or data access network, but they can also be transitory in nature. An application may experience errors in response to IMS procedures such as registration, Domain Name System (DNS) lookup failures, or it may simply be unable to access a data bearer. For each of these scenarios, the behavior of the IMS client needs to be defined with the goal of restoring service as soon as is practical.

Provisioning errors occur when the UE, ISIM, or network elements are not properly configured for a user to access desired services. These errors may result in an experience similar to the previously discussed error types, and likewise clients should also be able to recover automatically once provisioning problems are corrected.

Testing these error scenarios also poses challenges. Many types of errors and impairments cannot be readily recreated in a live environment. Long-term testing may eventually cause specific ones to be seen, but testing against simulators is the only practical method to verify error handling in the client. This adds significantly to the testing load as there are usually more failure cases to consider versus testing only for normal operation.

9.6 User Experience and Application Interactions

9.6.1 UE Hardware/Software Limitations

The user experience provided by an application is just as important as its technical design. The use of IMS as the core technology should not be considered the only factor in determining whether an application is going to be successful. It is the total user experience that matters, as this is what users will base their opinions upon. While PC-based applications have the luxury of extensive hardware capabilities, memory, and screen space, applications running on mobile devices face numerous limitations. These may impact the quality of the user experience, and must be factored into the client design so that appropriate compromises in user interface (UI) or functionality can be made:

- Multitasking capabilities: Due to processor or memory limitations, it may not be possible to run multiple applications simultaneously. If there is a need for multiple IMS applications to run concurrently (either to provide real-time updates or to share data between them), the UE must have a corresponding minimum set of capabilities. It is also important that applications running in the background not utilize services that generate charges without there being a benefit to the user. For example, it is appropriate for a user-requested file download to occur in the background; however, an interactive application should probably be suspended.
- Total bandwidth requirements: The capabilities of the access network may support the needs of one application, but not multiple ones simultaneously. In these cases, one application may need to be suspended while another is active.
- Display capabilities: Display size, resolution, and navigation capabilities may limit the user's ability to easily run multiple applications simultaneously or view large amounts of information.
- Device I/O capabilities: Mobile devices typically have one primary display, and in many cases, they can only output from one audio source at a time. Combined with limited processor power and bus bandwidth, this eliminates some options for simultaneous activities. For example, audio streaming will probably need to be paused to perform a voice call.

9.6.2 Data Sharing

Data sharing and interaction between device applications are a means of providing a richer end-user experience. When two or more applications on a UE need access to the same data, it is an inefficient use of both UE and network resources to duplicate enabler functionality in each one. By separating enabler data from the application, one can make this information available to any application on the UE (as opposed to building individual silo applications, which impedes data sharing). This is one of the benefits of using a common IMS framework. For example, current non-IMS Instant Messaging (IM) and Push-to-Talk over Cellular (PoC) applications are implemented as independent (i.e., silo) applications. Even though they may share mutual contacts, each application will independently query the network for the same presence data. Under our layered generic architecture, with the presence information abstracted out of the application, it can now be accessed by independent UE applications for consumption and sharing.

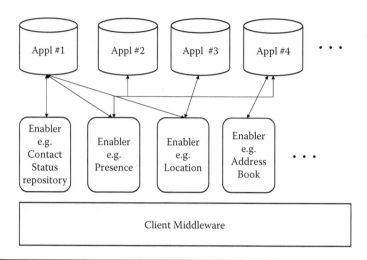

Figure 9.5   Generic application data sharing architecture.

Care must be taken in the data sharing design process, however. Due to the issues with mobile applications discussed previously, there may need to be a limit imposed on when sharing of data is allowed. Another consideration is that if applications cannot run simultaneously, the shared data may not always be up to date. Figure 9.5 shows a generic client architecture that includes sharing of data between IMS applications. This represents the full realization of the promises of IMS: multiple applications running on a common framework that allows the seamless sharing of data between them.

Endnotes

1. A Packet Data Protocol context is a 3GPP packet-switched connection over which the UE and the core network exchange IP packets.

Further Reading

Open Mobile Terminal Platform, "OMTP IMS Function Requirements," http://www.omtp. org.
"JSR 180: SIP API for J2ME," http://jcp.org/en/jsr/detail?id=180.
"JSR 281: IMS Services API," http://jcp.org/en/jsr/detail?id=281.
"JSR 325: IMS Communication Enablers (ICE)," http://jcp.org/en/jsr/detail?id=325.
OMA, "Device Management," version 2.1, http://www.openmobilealliance.org.
OMA, "Enabler Release Definition for Client Provisioning," version 1.1, http://www.open-mobilealliance.org.

3GPP TS 24.229, "IP Multimedia Call Control Protocol Based on Session Initiation Protocol (SIP) and Session Description Protocol (SDP); Stage 3."

3GPP TS 33.203, "3G Security; Access Security for IP-Based Services."

3GPP TS 33.978, "Security Aspects of Early IP Multimedia Subsystem (IMS)."

3GPP TS 31.103, "Characteristics of the IP Multimedia Services Identity Module (ISIM) Application."

RFC 3344, "IP Mobility Support for IPv4," Perkins, C., Ed., August 2002.

RFC 3775, "Mobility Support in IPv6," Johnson, D., Perkins, C., and J. Arkko, June 2004.

RFC 2205, "Resource ReSerVation Protocol (RSVP)," Braden, R., Zhang, L., Berson, S., Herzog, S., and S. Jamin, September 1997.

IMS SIGNALING PRIMER

Chapter 10

What's in a Name
Identifiers in IP Networks

Session Initiation Protocol (SIP) is the linchpin signaling protocol in IP Multimedia Subsystem (IMS). Accordingly, SIP addresses are the key identifiers for IMS network subscribers.

However, routing at the Internet Protocol (IP) layer is based on IP addresses. So, in a sense, the endgame always entails getting to an IP address: the most basic type of identifier in an IP network. Hence, before moving on to discuss SIP addresses and other higher-layer identifiers, we briefly discuss IP addresses.

There are three optional sections toward the end of this chapter: Section 10.7 (on IP addresses), Section 10.8 (on some subtleties of SIP addresses), and Section 10.9 (on Network Access Identifiers). These sections provide some additional technical detail for the interested reader. The rest of this book does not depend on the optional sections, so the reader can skip those sections, if desired, without reservation.

10.1 IP Addresses

IP addresses—192.168.0.1 is an example—are strings of digits. End users can be insulated from dealing directly with IP addresses by presenting text-string identifiers to humans and letting machines keep track of the mappings between those identifiers and IP addresses. This is convenient for a variety of reasons, including the following:

- Long strings of digits are not mnemonic. Having end users enter "raw" IP addresses would be error prone.

- IP addresses are often dynamically allocated. Even in the case of so-called static IP addresses, network administrators may occasionally have to perform significant reconfigurations of IP address assignments.
- Reliability, load balancing, and other requirements often dictate that multiple machines (having different IP addresses) be associated with the same text-string identifier. For example, popular Web portals distribute traffic among large numbers of Web servers. This is transparent to end users who access Web portals.

10.2 Uniform Resource Identifiers

SIP addresses are Uniform Resource Identifiers (URIs). URI is a generalization of the more familiar term Uniform Resource Locator (URL). We will not be strict about the distinction between URIs and URLs. Moreover, the formal syntax specification for URIs makes for dull reading. We will dispense with formalities and work instead with examples:

- `mailto:leo@tolstoy.com` is a mailto URI (aka, an e-mail address).
- `http://www.example-website.com/~leotolstoy` is a Hypertext Transfer Protocol (HTTP) URI. (HTTP is the key protocol for accessing the World Wide Web.)
- `tel:+1-512-555-1212` is a tel URI.
- `sip:bob@biloxi.com` is a SIP URI.

For e-mail addresses, the mailto: prefix is often omitted. E-mail is so familiar that people assume by default that a string of the form `leo@tolstoy.com` is, in fact, an e-mail address. Moreover, e-mail programs do not require users to type in the mailto: prefix. Similarly, if a user types `www.example-website.com/~leotolstoy` into a Web browser, the browser supplies the http:// prefix.

Here is something to think about: why is there an "@" sign in the mailto and sip URIs, but not in the http and tel URIs? We will answer this question in Chapter 11, Section 11.5.

10.3 SIP URIs

As shown in Figure 10.1, a SIP URI can be decomposed into three parts—the URI scheme identifier (which is simply "sip"), the user part, and the domain part. The delimiter between the URI scheme identifier and the user part is always ":", and the delimiter between the user part and the domain part is always @. The term *username* is sometimes employed instead of *user part*.

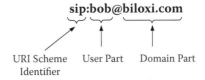

Figure 10.1 Anatomy of a SIP URI.

Strictly speaking, SIP URIs do not always have user parts. As an example, `sip:server1.att.net` could be used as a URI for a SIP server.

In keeping with the de facto industry parlance, we use the terms *SIP Address* and *SIP URI* interchangeably. Not all SIP addresses are created equal. Next, we draw an important distinction between two types of SIP addresses.

10.3.1 SIP Addresses of Record

The SIP URI we have seen already—`sip:bob@biloxi.com`—is an example of a SIP *Address of Record (AoR)*. A distinguishing feature of this type of address is that its domain part is not very specific, in the following sense: Bob's AoR tells us that he receives SIP services through the biloxi.com domain, but nothing else. In particular, the AoR does not identify a device (e.g., a landline phone, a PC, or a cellular phone) where Bob might be available. Thus, the information in the AoR is not sufficient to actually route SIP messages to Bob.

10.3.2 SIP Contact Addresses

There is a distinct notion of SIP contact address, in which the domain portion identifies a host device. In practice, the domain portion is usually an IP address. (Recall that the domain portion of a SIP URI is the part after the @ sign.) Suppose Bob has a SIP phone sitting in his office whose IP address is `192.168.0.1`. Then `sip:bob@192.168.0.1` could serve as a contact address for Bob. From the reader's point of view, it is cumbersome to have IP addresses sprinkled into the exposition. We will therefore substitute `sip:bob@<desk-ip-address>` for the contact address shown earlier. We emphasize that the latter version of the contact address is not to be taken literally, as it is not a legal SIP URI. Rather, the string <desk-ip-address> should be taken to mean "the IP address of Bob's desk phone goes here."

To reach Bob, it will ultimately be necessary to obtain a contact address. However, an AoR is still a very useful thing, because it takes into account the possibility that Bob can be reached at a variety of devices connected to the biloxi.com domain. Suppose, for example, that Bob is sometimes best reached via his SIP-enabled mobile phone. On such occasions, perhaps his contact address is `sip:bob@<cell-ip-address>`.

Summing up, the AoR is an appropriate address to publish (e.g., on a business card), as it insulates people who want to reach Bob from having to keep track of his varying contact addresses.

The foregoing discussion implies that the network must somehow keep track of dynamic mappings between AoRs and contact addresses. How is this accomplished? How, when, and where does the network "de-reference" AoRs to contact addresses (i.e., what are the procedures for substituting contact addresses for AoRs in SIP messages)? We will examine these questions in the SIP overview chapter.

10.4 Tel URIs

It is fair to say that SIP URIs are the "native" identifiers in IMS networks. However, telephone numbers (TNs) are also important identifiers. This holds true not only in traditional circuit-switched networks but also in IMS networks. We can cite a number of reasons for the continued importance of TNs:

■ Familiarity. End users are familiar with TNs. The following observations reinforce this point:
  – IP telephony is on its way to becoming a widely deployed IMS service. Using TNs, as in the past, makes IP telephony seem less like a strange new service.
  – In the cellular world, TNs are familiar identifiers for data services (i.e., Short Message Service and Multimedia Message Service) as well as traditional voice services.

■ Routability in the circuit-switched domain. From the addressing point of view, this is the key to interoperability between IMS and the circuit-switched domain. Being able to reach (and be reached by) circuit-switched endpoints is clearly a must-have for IP telephony.

■ Usability. It is much easier to type a TN into a telephone keypad than it is to type a SIP URI into a telephone keypad.

■ TNs are routinely distributed contact information. It is a nice added convenience if these same identifiers can also be used for peer-to-peer data services (rather than having to keep track of separate identifiers).

■ Use of TNs facilitates services that combine circuit-switched voice calls with concurrent data sessions.

Clearly, there is motivation to allow the use of TNs within IMS networks. To do so, TNs must be presented as tel URIs. The string "tel:" alerts the IMS network that it is looking at a TN rather than a SIP URI. By way of example, here are two tel URIs:

```
tel:+1-512-555-1212
tel:555-1212; phone-context=+1-512
```

These two tel URIs represent the same phone number. The differing representations could arise for a variety of reasons. One possibility is that, in the second URI, 555-1212 is the so-called *dial string* (i.e., the set of digits that the caller explicitly dials). Perhaps, 7-digit dialing has always worked for this number in the past. However, because this 7-digit number is not globally unique, a switch or gateway appends the phone-context parameter. Taken together, the dial string and phone-context parameters specify a globally unique TN. A caller who is trying to reach the same number from overseas will not be able to do so with 7-digit dialing, but instead must enter the globally unique TN explicitly, giving rise to the first of the two URIs.

The notions of SIP AoR and contact address are not specific to IMS. In fact, these concepts are covered in the base SIP specification. The address types introduced in the following section are specific to IMS. IMS identities are often, but not always, SIP URIs.

10.5 IMS Identities

10.5.1 Private User Identities

IMS Private User Identities (IMPIs) take the form `username@realm`. They are utilized for Authorization, Authentication, and Accounting purposes, and their heaviest use is during the process of registering with an IMS network. End users are usually not aware of the IMPIs associated with their subscriptions.

Given their role in authentication, it is important that IMPIs be kept secure. This need has given rise to fraud-prevention measures governing their usage. For example, IMPIs cannot be altered by end users and are not used for SIP routing. In fact, Private User Identities are not formatted according to the SIP URI scheme (they lack the "sip:" prefix). Moreover, IMPIs are exposed outside the home IMS network as little as possible (although roaming scenarios dictate that they must be exchanged by visited and home networks during registration).

In Universal Mobile Telecommunications System (UMTS™) networks, IMPIs are based on International Mobile Subscriber Identities (IMSIs); the same is true of GPRS and EDGE networks. (GPRS and EDGE are precursors of UMTS.) The username is equal to the International Mobile Subscriber Identity, which is a 20-digit identifier housed on a smartcard. The realm portion of the IMPI is constructed according to a set formula. "Filling in" this formula requires that two subfields be copied from the International Mobile Subscriber Identity: namely, the Mobile Network Code and Mobile Country Code.

10.5.2 Public User Identities

IMS Public User Identities (PUIDs)[1] can be SIP URIs or tel URIs. The motivation for allowing tel URIs has been documented in Section 10.4. However, tel URIs are not routable in SIP domains. Therefore, the IMS specifications state that, in order to route a SIP message to an endpoint identified by a tel URI, the tel URI must first be mapped to a "SIP routable SIP URI."

An IMS subscriber may have multiple PUIDs. Every IMS subscriber must have at least one PUID that takes the form of a SIP URI. This type of PUID can be regarded as a SIP AoR. An IMS subscriber may optionally have one or more PUIDs that take the form of tel URIs.

The IMS specifications allow for multiple contact addresses to be associated with the same PUID at the same time. Associating one contact address with multiple PUIDs is also allowed. This is sometimes summarized by saying that there is a *many-to-many* relationship between PUIDs and contact addresses.

10.5.3 Public and Private Service Identities

Typically, we think of IMS URIs as representing end users (i.e., humans). Public Service Identities were conceived to accommodate other kinds of SIP endpoints. Perhaps the best example is a chat room, which would be assigned a SIP URI. After all, the chat room does need to be an addressable entity. However, the chat room would not be associated with any individual.

Continuing with our example, there are certain tasks that the chat room itself has to take care of (e.g., managing logins and logouts as users come and go; broadcasting incoming messages to the set of users who are currently logged in). The software processes that accomplish these tasks are running on a server that may or may not reside in the IMS service provider's network. If the server and the IMS core elements are in different domains, the chat room server will have to be authenticated before it can connect to the IMS network. In such cases, there is a need for a secure identifier. To fill this niche, the IMS specifications define Private Service Identities. Similar to Private User Identities, Private Service Identifiers take the form of Network Access Identifiers. Beyond this requirement, however, each network operator is free to define its own format for Private Service Identities.

10.6 Mappings Among Identifiers in IP Networks: DNS and ENUM

Recall, from the beginning of this chapter, that routing in IP networks is based on IP addresses. However, to make life easier for end users as well as network administrators, it is worthwhile to utilize text-string identifiers that are mapped to IP addresses. The scheme for maintaining and resolving those mappings is called the Domain Name System (DNS). DNS functions as a database in which the records are dispersed among numerous geographically diverse locations.

So far, we have said that DNS can be used to map text strings to IP addresses. So, specifying such a text string (we will not worry about the syntax) is an indirect way of referring to an IP address. Over the years, additional layers of indirection have been added to the original DNS functionality. An early example of this has to do with mail servers. If the domain portions of e-mail addresses (e.g., hotmail.com or mail.yahoo.com) mapped directly to IP addresses, that would make it difficult for large Internet service providers (ISPs) to manage their server farms. Instead, the domain portion of an e-mail address typically maps to text strings identifying servers, and those strings in turn map to IP addresses.

In fact, the process of resolving a text-string identifier to an IP address can require several steps, that is, following a sequence of DNS mappings until finally arriving at an IP address. In general, the details are not important to us. However, one particular DNS application merits special mention here: ENUM. The term ENUM, which can be expanded as Electronic NUMbering or tElephone NUMber mapping. refers to mapping, of telephone numbers to URIs. Formally speaking, IMS utilizes ENUM to map tel URIs to SIP URIs.

10.7 Optional Section: IP Addresses

The example IP address given earlier in this chapter—192.168.0.1—is an IP version 4 (IPv4) address. IPv4 predominates now, but will give way to IPv6 in the future. This is largely due to exhaustion of IPv4 address space. Each IPv6 address is 128 bits long, as opposed to 32 bits for IPv4. (Each of the quantities 192, 168, 0, and 1 in the sample IPv4 address is the decimal representation of an 8-bit field.) In the following example IPv6 address, each letter represents a hexadecimal digit (which is the equivalent of 4 bits):

FF01:0000:0000:0080:0700:1234:ABC0:0043.

Note also that the delimiter between groups of digits is ":", as opposed to "." for IPv4.

Some IP addresses are globally routable, whereas others are not. The example IPv4 address used earlier—192.168.0.1—is an instance of the latter. Certain IP addresses, including this one, have been set aside for use in private networks. At any given time, such an address may be simultaneously assigned to many end devices, each residing in a different private network. For example, the IPv4 address 192.168.0.1 sits in a range that is commonly assigned to PCs in home or small office networks. The assignment is most commonly performed by a router/firewall device.

The reusability described in the previous paragraph is beneficial from the address exhaustion point of view, but the resulting non-unique association (i.e., between a single address and multiple end systems) precludes any kind of global routability.

10.8 Optional Section: SIP Contact Addresses and Globally Routable User Agent URIs (GRUUs)

Recall from Section 10.3.2 that SIP AoRs are not generally sufficient to route SIP messages to end devices. SIP AoRs must be dereferenced to SIP contact addresses somewhere along the way. For completeness, we note here that the notion of contact address—as defined in the base SIP specification—is not sufficient to ensure that the aforementioned dereferencing works properly in all cases. As a result, the concept of Globally Routable User Agent URI (GRUU) has been developed. The reader can think of GRUUs as contact addresses that are robust in scenarios where SIP messages must cross multiple domains with differing IP routing logic. Rather than cover this subject in depth, we give an example that hints at its complexity.

There are use cases in which it is necessary to communicate with a specific SIP device. Imagine, for instance, that Bob is in an extended session with another person, whom we will call Alice. This is the same Bob who appeared in Section 10.3. Suppose the session is a voice call that Bob has chosen to take at his desk phone, and moreover, suppose that the desk phone can participate in interactive data sessions. Alice wants to send a file to Bob, and wants to make sure it reaches the phone he is currently using. If Alice sends a data session setup request to sip:bob@ biloxi.com (recall that this is Bob's SIP AoR), it may not be routed to his desk phone. On the surface, it seems that Alice should simply send the setup request to sip:bob@<desk-ip-address>—and in many cases this would indeed work. However, suppose that Bob's desk phone has a private IP address: specifically, an address that is not routable outside the biloxi.com domain. If Alice's phone resides outside the biloxi.com domain, the session setup request will not be routable and, therefore, the attempt will fail.

The idea of a GRUU is that a SIP protocol entity (ostensibly at the boundary of the biloxi.com domain) will assign a SIP URI that "makes sense in the outside world." This boundary protocol entity will establish—and keep track of—an association between the assigned URI and Bob's desk phone.

10.9 Optional Section: Network Access Identifiers

IMS Private User Identities follow the IETF specification for Network Access Identifiers (NAIs). In many cases, NAIs take the form <user part>@<domain part>. Except for the lack of a URI scheme identifier, this looks like a URI—and this is a reasonable way to think of NAIs for the remainder of this book. However, the syntax for NAIs must take into account cases in which one domain is accessed through another domain. This gives rise to more complex NAIs, such as `biloxi.com!bob@tolstoy.com`. The last NAI applies to scenarios in which Bob accesses his home domain, which is biloxi.com, through the tolstoy.com domain.

Endnote

1. It should be noted to avoid confusion that 3GPP is inconsistent in its acronym for "IMS Public User Identity." In other specifications (especially those related to security), the acronym "IMPU" is used.

Further Reading

RFC 791/STD 0005, "Internet Protocol," IETF, September 1981.
RFC 2474, "Definition of the Differentiated Services Field (DS Field) in the IPv4 and IPv6 Headers," IETF, December 1998.
RFC 2460, "Internet Protocol, Version 6 (IPv6) Specification," IETF, December 1998.
RFC 4282, "The Network Access Identifier," IETF, December 2005.

Chapter 11

Overview of Session Initiation Protocol (SIP)

11.1 What is the Purpose of SIP?

SIP [1] is used to control interactive peer-to-peer sessions. From a telco point of view, the classic example of an interactive session is a simple phone call. In generic terms, the following steps are required to set up a call. This is true regardless of whether the call is placed across a packet-switched network, for example, using Voice-over-IP (VoIP), or a traditional circuit-switched network. Moreover, the same basic steps are equally applicable to video and multimedia sessions.

> **Step 1.** A session setup request is issued by the calling party.
> **Step 2.** The called party indicates that it has received the request and is alerting the end user.
> **Step 3.** The called party confirms that it is willing to engage in a call.
> **Step 4.** Details regarding the *bearer path* are confirmed. (For a definition of bearer path, read on to the end of this paragraph.)

In this section, we will cast these generic steps in terms of SIP. We will also cover session teardown procedures, although they are omitted from the steps listed above. First, some comments are in order. The reader should view the calling party and the called party as software processes carrying out the wishes of end users (rather than the end users themselves). The steps listed above take place in the *Control Plane*.

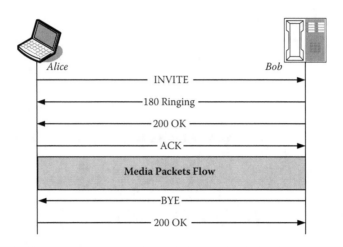

Figure 11.1   Simple SIP session setup and teardown.

Voice—and/or video, as the case may be—is encoded and transmitted across the *Bearer Plane*. The control and bearer planes have separate functions and are logically distinct. For at least part of the transmission path, the two planes are usually physically distinct as well. The bearer path is simply the transmission path through the bearer plane.

Increasingly, VoIP is being deployed as a SIP-based service. Starting with the next paragraph, we walk through a VoIP call flow using SIP as the call control protocol. However, SIP was designed from the beginning to be flexible, and should not be regarded purely as a VoIP call control protocol. In fact, SIP is applied to a wide array of services. Many such services—but not all of them—are real-time services.

In the classic example Alice indicates (say, via the user interface [UI] on her PC) that she wants to place a call to Bob (Figure 11.1). The treatment given here is loosely based on Section 4 of [1]. To set up the call, the two devices (that is, Alice's PC and Bob's phone) exchange SIP messages as follows. The SIP software on Alice's PC sends a SIP INVITE message to Bob's desk phone. Bob's phone rings (as indicated to the far end by the SIP 180 Ringing message); when Bob responds by accepting the incoming call, his phone sends a SIP 200 OK message to Alice, who replies with a SIP ACK message (acknowledging receipt of the 200 OK). The VoIP session setup is now complete, and Alice and Bob's conversation can proceed. Referring back to the generic steps at the beginning of this section, we have the following: the INVITE message implements step 1, the 180 Ringing message implements step 2, the 200 OK message implements step 3, and the ACK message takes care of step 4.

When the call is over, the session has to be torn down. At the end of the conversation, suppose that Bob hangs up first. In this case, his phone sends a SIP BYE and receives a SIP 200 OK in response. The teardown process was not covered at the beginning of this section, so the BYE/200 OK exchange does not correspond to any of the generic steps listed there.

In the example given here, Alice and Bob conduct a voice conversation. But if this had been a video or Instant Messaging session instead, the picture would look exactly the same. Thus, we have purposely annotated the appropriate part of the figure with the phrase "Media Packets Flow" instead of "Voice Packets Flow." The parameters within the SIP messages will, of course, vary depending on the type of session involved.

Another point is that the media packets are not SIP packets. SIP is used to negotiate the parameters for the session, paving the way for another protocol to do the job. The session parameters are carried in the bodies of SIP messages using Session Description Protocol. Parameter negotiation is handled according to the Offer/Answer Model. Session Description Protocol and the Offer/Answer Model are covered in Chapter 12.

11.2 Routing of SIP Messages

SIP is purpose-built to facilitate person-to-person interaction. It can also be used to control communication between persons and interactive applications (e.g., interactive voice response systems).

In the interest of simplicity, we suppressed an important detail in Section 11.1: how does Alice find Bob? Or, more precisely, how is address resolution accomplished within Alice's and Bob's networks? A truly comprehensive answer to this question is beyond our scope, as address resolution takes place in numerous stages. However, the following limited discussion of SIP routing and address resolution will pave the way for later sections in this chapter.

The example in Section 11.2 depicts signaling between SIP endpoints—that is, Alice's PC and Bob's desk phone—without any intermediaries. Practically speaking, SIP servers are necessary for successful call routing. In Figure 11.2, we revisit the scenario of Figure 11.1 with additional detail. The first thing to notice is the presence of SIP URIs for Alice (sip:alice@<pc-ip-address>) and Bob (sip:bob@<desk-ip-address>). The reader should think of these as contact addresses corresponding to Alice's and Bob's Addresses of Record (AoRs), respectively. In the contact address for Bob, <desk-ip-address> is a substitute for the IP address of Bob's desk phone (and similarly for Alice's PC). We have performed this substitution in the interest of readability. This is the same approach we adopted in Chapter 10, Section 10.3.2.

As we continue this example, we will use sip:alice@att.net as Alice's AoR and sip:bob@biloxi.com as Bob's AoR. AoRs, rather than contact addresses, would normally be given out as contact information. That is what we assume in this example.

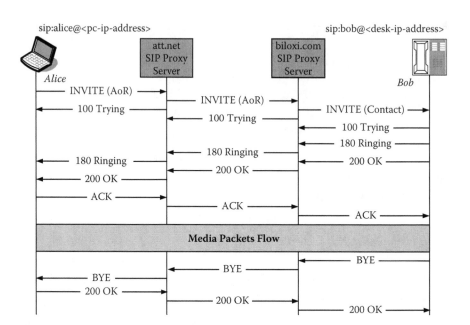

Figure 11.2 A more realistic depiction of a SIP session.

Figure 11.2 also adds SIP servers in Alice's and Bob's domains. These are called proxy servers because they act on SIP endpoints' behalf (e.g., by forwarding call control signaling as in this example). When Alice indicates that she wants to call Bob, her PC sends an INVITE toward Bob's AoR. Instead of trying to locate Bob directly, Alice's PC sends the INVITE to the SIP proxy for the att.net domain. Here the domain part of Bob's AoR is important; that piece of information is used by the att.net proxy to locate the biloxi.com proxy. Once it has done so, the att.net proxy can forward the SIP INVITE. Note that the labeling of the first two INVITEs in Figure 11.2 indicates that, so far, the INVITE is directed to Bob's AoR.

In turn, the biloxi.com proxy knows that the best place to reach Bob is at his desk phone and forwards the INVITE thereto. Note the labeling of the third INVITE in the figure, which indicates that the AoR has been resolved to a contact address.

The 180 Ringing and 200 OK follow the same path in reverse, and so on. (Following the same path in reverse is typical, but strictly speaking, it is not mandatory.)

The media packets do not have to flow through the SIP servers in the diagram. Using the terminology introduced in Section 11.1, the SIP servers do not reside in the bearer plane.

Note that a real-world signaling flow would typically include additional messages that, for simplicity, are not shown in Figure 11.2. SIP protocol entities have timers (e.g., so they can retry a SIP INVITE if a certain amount of time passes without receiving a response). The `att.net` proxy would normally send a SIP `100 Trying` message to Alice's PC upon receipt of the `INVITE`. This lets Alice's PC know that the `INVITE` has reached the "next hop" and timers can be set accordingly. Similarly, upon receiving the forwarded `INVITE`, the `biloxi.com` proxy would send a `100 Trying` message to its `att.net` counterpart. However, the `100 Trying` message is not forwarded in either case.

In the signaling flow we have just described, how does the `biloxi.com` proxy server know it should forward the `INVITE` to Bob's desk phone? Moreover, how does Alice's PC find the `att.net` proxy server in the first place? We will return to these questions in Section 11.8.

11.3 SIP Forking

In this section we cover a simultaneous ringing scenario. As before, Alice calls Bob. However, in this case, Bob wants his desk phone and his wireless phone to ring simultaneously. If he is away from his desk, he will answer the call on his wireless phone. If he is at his desk, he would prefer to answer the call on his desk phone. Of course, once Bob answers the call on either phone, he wants both phones to stop ringing.

SIP is expressly designed to accommodate such scenarios through a feature called forking. SIP forking is illustrated schematically in Figure 11.3.

Alice sends a SIP `INVITE` to Bob's AoR. In this example, the SIP Proxy Server has two contact addresses for this AoR, and it forwards the SIP `INVITE` to both

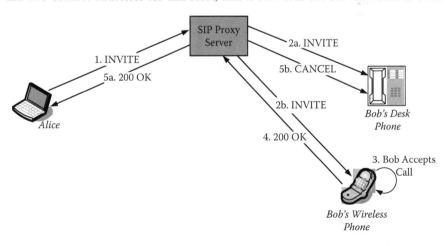

Figure 11.3 Simultaneous ringing scenario implemented with SIP forking.

of them in steps 2a and 2b. In step 3, Bob accepts the call on his wireless phone, resulting in a 200 OK response in step 4. The SIP proxy then does the following:

- Forwards the 200 OK to Alice's PC (as in the previous scenario; this is shown as step 5a).
- Sends a SIP CANCEL message to Bob's desk phone (to tell it to stop ringing; this is new to the current scenario and is shown as step 5b).

In the interest of simplicity, some of the details that appeared in the flow of Section 11.2 are omitted here. For example, 180 Ringing messages would be sent by Bob's desk and wireless phones upon receipt of Alice's INVITE. Moreover, Alice's PC will issue an ACK upon receipt of the 200 OK (in step 5a).

The forking capability we have described here gives SIP a great deal of flexibility. It can also be used to implement *sequential* ringing scenarios (e.g., ring Bob's desk phone first; if there is no answer, try his wireless phone).

11.4 Anatomy of SIP Messages

The att.net and biloxi.com proxy servers decide what to do by looking at the *header* fields in the SIP messages they receive. (This is also true of the SIP software on Alice's PC and Bob's desk phone, although we are more interested in proxies in this section.)

Every SIP message is a request (which begins with a request-line) or a response (which begins with a status-line). The first line ends with a carriage return/line feed sequence (CRLF) and is followed by a sequence of headers. Headers are separated from one another by a CRLF. So, when a SIP message appears on the printed page, it will take up multiple lines. In the example of Section 11.2, the first several lines of Alice's initial INVITE look like this:

```
INVITE sip:bob@biloxi.com SIP/2.0
Via: SIP/2.0/UDP pc.att.net;branch=qK19035809
Max-Forwards: 70
To: Bob <sip:bob@biloxi.com>
From: Alice <sip:alice@att.net>;tag=42398752
<remainder of SIP message has been truncated>
```

Note that Bob's AoR appears on the first line, right after the INVITE. The via header in the second line says that the sender can be reached at pc.att.net. In the same line, UDP pertains to the layer below SIP and is beyond our scope. The branch parameter is also beyond our scope, although we remark that it "lives" at the SIP layer. Later in the INVITE, there is a From header that identifies Alice as the sender. Since the Via header does not identify an end user, and in light of the fact that Alice may not be the only person who uses this PC, the From header is clearly not redundant. Notice the tag parameter at the end of the From header. This parameter value is machine generated (as is the branch parameter mentioned) and its purpose is to make sure that the current signaling interchange, or dialog, can be uniquely identified. This is important because many SIP dialogs can be in progress simultaneously.

We have not looked at the To header yet; let us compare it to the From header. Note the presence of the *display names* Alice and Bob (before the "<" character delimiting the beginning of the SIP URI). The biggest difference is that the To header lacks a tag parameter. When Bob's phone issues a response (e.g., the 200 OK in our example), the tag field will be populated. The From tag that was generated by Alice's PC will be copied unchanged into the 200 OK. In the 200 OK, the To field will still identify Bob (and the From field will still identify Alice), even though the message is traveling in the other direction.

We return to the INVITE that was issued by Alice's PC. What happens to this request as it progresses toward Bob's phone? The att.net proxy adds a Via header indicating its presence in the message path before forwarding the INVITE, as does the biloxi.com proxy. Moreover, the Max-Forwards header is decremented to 69 by the att.net proxy and then to 68 by the biloxi.com proxy. (Note: The SIP specification [1] recommends an initial Max-Forwards value of 70, so we have used that value in this example.) Thus, the collection of header fields in a SIP message can change en route to its destination. The baseline SIP specification [1] defines some 44 header types, and additional header definitions can be found in other IETF specifications. Detailed information about SIP headers is beyond our scope; suffice it to say that the variety of available headers afford great flexibility and power in designing SIP-based applications.

The SIP message *body* comes after the headers and is separated from the headers by two consecutive carriage return/line feed sequences. Thus, on the printed page, a blank line will appear between the last header and the beginning of the message body. We will give an example in Section 11.11.6. Note also that Chapter 12 examines message bodies used for session parameter negotiation in detail. The message body is not mandatory, and in fact, SIP messages will sometimes have empty bodies.

The main point we want to make here is as follows: the proxy servers between Alice and Bob do not care about SIP message bodies. They look to the message headers for guidance in performing their tasks. This is not to say that proxy servers are prohibited from looking at SIP message bodies, but only that they normally do not need to do so. This allows for a high degree of independence between message headers and message bodies, which fulfills a design principle that the creators of SIP had in mind from the outset.

We now summarize the content of this section. The first line of each SIP message is either a request-line or a status-line, according to whether the message is a request or a response. SIP requests are covered in Section 11.11, and responses are covered in Section 11.12. Following the first line, there is always a sequence of headers. Headers are separated from the first line and from one another by CRLF. In other words, each header occupies a separate line. The sequence of headers is separated from the message body by two consecutive CRLFs. SIP headers can contain parameters (e.g., the branch and tag parameters encountered in the example of this section). In this book, we will discuss headers from time to time as needed, but we will not attempt anything approaching comprehensive coverage. We will discuss parameters sparingly. For a detailed treatment of headers and parameters, the reader can consult [2].

11.5 Comparing and Contrasting SIP and HTTP

Hypertext Transfer Protocol (HTTP) [3] is the predominant protocol for accessing the World Wide Web. As the profusion of useful Web applications attests, HTTP has been extremely successful in achieving the goals set by its designers. The developers of the SIP protocol borrowed heavily from the design of HTTP. Both protocols are text based. That is, SIP and HTTP messages are human readable, although some parts will make more sense to the nonexpert reader than others.

As we go through this chapter, we will point out other similarities between SIP and HTTP. The differences between the protocols are also telling, so we will highlight some of those differences as well.

As noted in the previous section, SIP is oriented toward peer-to-peer communication. Unlike SIP, HTTP is tailored to a client-server paradigm. In the quintessential World Wide Web experience, end users interact with Web servers rather than other end users—HTTP is used to download content from (and upload content to) Web servers. Summing up, there is a difference between the underlying paradigms for SIP and HTTP. In fact, the difference in emphasis between the two protocols is paralleled by a difference in the URI schemes. Recall that HTTP URIs lack user parts, as reflected by the lack of an "@" sign. The example HTTP URI from the previous chapter, namely, `http://www.example-website.com/~leotolstoy`, will serve to jog the reader's memory. Contrast this HTTP URI with the SIP URI `sip:bob@biloxi.com`. When working in the context of SIP, it is important to be able to identify end users.

HTTP and SIP are similar in that both protocols are designed to make use of proxy servers as routing intermediaries. Both protocols make heavy use of header fields to carry information that affects routing. Header fields also serve other purposes, but the details are beyond our scope.

11.6 SIP Application Servers

Although the example of Section 11.2 is more realistic than its predecessor in Section 11.1, it still is oversimplified in one crucial regard: it does not include an application server. Most services will require implementation of algorithms—which we will refer to as *service logic*—that take a variety of scenarios into account and act appropriately. Service logic resides on SIP application servers. In our earlier examples, Bob is available to take Alice's call. If we were to consider call forwarding behavior for instances when Bob is away from his phone (e.g., forward to cell phone or to voice mail), one or more application servers would normally be used to implement the necessary service logic.

11.7 SIP Functional Entities

SIP functional entities include user agents (UAs), registrars, proxy servers, redirect servers, and back-to-back user agents (B2BUAs). To avoid confusion while reading the balance of this chapter, it may be helpful to keep the following points in mind:

- A single network element might incorporate more than one SIP functional entity.
- SIP application servers are purposely not listed as SIP functional entities. The role played by an application server can differ depending on circumstances. A common example is that an application server would act as a proxy server in some cases and as a UA in other cases.

11.7.1 User Agent (UA)

A SIP UA is most commonly a piece of software running on an end user's terminal (e.g., Alice's PC or Bob's desk phone) or a software instance running on a SIP application server. End users' terminals can take a variety of forms; examples include PCs, SIP phones, residential gateways, cellular phones, and IPTV set top boxes.

One can think of UAs as SIP *endpoints*. This is in the sense that the ultimate recipient of a SIP message (after routing through intermediaries such as SIP proxies) is usually a UA.

11.7.2 Registrar

This element keeps track of subscribers' current status. More specifically, a registrar manipulates (i.e., creates, modifies, deletes) bindings between SIP Addresses-of-Record and Contact Addresses. The database that stores the bindings can be resident in the same network element that houses the registrar. Alternatively, as is the case with IMS, the database can be separate and can be accessed by the registrar via some other protocol. From the SIP point of view, it does not matter.

For HTTP, there is no concept of a registrar. This makes sense: because we are safe in assuming that a Web server for a popular Web portal will not leave the office, we can be confident that we will not need to reach that server on its cell phone. Thus there is little need for a distinction between HTTP Address of Record and HTTP contact address.

11.7.3 Proxy Server

SIP proxy servers—or simply *SIP proxies*—are signaling intermediaries. We have already seen that SIP proxies forward SIP messages (e.g., INVITEs and responses to INVITEs) on behalf of UAs. It is not uncommon for a single network element to function as a registrar and also as a proxy.

SIP proxies can be stateful or stateless. Throughout the duration of each signaling exchange, a stateful proxy retains information about the messages that have passed through it during that exchange. If, for example, the att.net proxy in Figure 11.2 is stateful, it will recognize the INVITE, 200 OK, and ACK messages as pertaining to the same session setup procedure. A stateless proxy forwards each message based on the message headers and then "forgets" everything about that message.

The use of the proxy as a major functional component is one of the strongest similarities between SIP and HTTP. In each case, proxy servers comprise a routing layer that sits above the IP layer.

11.7.4 Redirect Server

A redirect server responds to SIP requests, indicating to the senders of those requests that they should contact alternate URIs—that is, URIs that are different from those to which the original SIP requests were addressed. In other words, a redirect server is just what it sounds like.

11.7.5 Back-to-Back User Agent

Perhaps the concept of back-to-back user agent (B2BUA) is best put across by example. In Sections 11.1 and 11.2, we have tacitly assumed that Alice and Bob know each other. In some use cases, however, Alice may want to place a call to a stranger without revealing her identity. Suppose, for instance, that Alice works in a customer service call center and that service representatives' identities are concealed from the customers who call in for security reasons. If Bob is a call-in customer, a B2BUA could be employed to mask Alice's identity from him. In terms of Figure 11.2, the att.net server would act as a B2BUA rather than a proxy server.

Alice's PC could still send the initial INVITE with a "From:" address of sip:alice@att.net. But perhaps the INVITE that reaches Bob shows sip:agent99@att.net as the originator of the call. The B2BUA must keep track of the fact that both SIP URIs represent Alice, so the identity-hiding function provided here comes at a cost in terms of processing and memory resources.

In this discussion, we are assuming that agent99 is a temporary alias for Alice—tomorrow, agent99 might be someone else—and that the att.net server manages this association dynamically. Alice's SIP UA does not have to be aware that the outside world is seeing an alias—that UA continues to think of itself as sip:alice@att.net. Later, Alice's employer may need to know that Alice was the representative who took Bob's call. Thus, it may be necessary for the att.net server to perform record-keeping duties so that this information can be recovered.

11.8 Server Discovery and Registration

Before the signaling flow of Figure 11.2 can take place, certain precursor steps must be carried out. Alice's PC must locate a SIP server in the att.net domain, and Bob's SIP phone must do likewise in the biloxi.com domain. Typically, each SIP endpoint will obtain a server hostname (either via static provisioning, or dynamically via signaling). Using that hostname, the SIP endpoint then obtains the server's IP address via lower-layer signaling. The same thing happens with HTTP: when a PC Web browser first opens, often a message saying "detecting proxy settings" (or suchlike) will be displayed. Moreover, Web browsers allow users to specify an HTTP proxy.

Next, each SIP endpoint must register with a SIP server. Among other things, this registration process is the means by which the biloxi.com server comes to know the current IP address of Bob's SIP phone.

Figure 11.4 shows a registration-signaling interchange. The first several lines of the initial REGISTER request might look like this:

```
REGISTER sip:registrar.biloxi.com SIP/2.0
Via: SIP/2.0/UDP desk.biloxi.com;branch=arF23095kk
Max-Forwards: 70
To: <sip:bob@biloxi.com>
From: <sip:bob@biloxi.com>;tag=257n932487
Contact: <sip:bob@desk.biloxi.com>;expires=3600
<remainder of SIP message has been truncated>
```

The SIP URI on the first line of the REGISTER request identifies the registrar for the biloxi.com domain. The lack of an "@" sign is due to the fact that the URI identifies a functional element rather than a specific SIP endpoint. The Via and Max-Forwards headers serve the same purpose as in Figure 11.2. Next, note that the To and From headers are both populated with Bob's AoR. This REGISTER request asks that the contact URI (i.e., the URI given in the Contact header) be bound to Bob's AoR. In most real-life cases, the portion of the Contact URI after the "@" sign will contain an IP address rather than a domain name. The expires parameter in the Contact header represents a request by the UA that the registration will endure for 3600 seconds—which equals 1 hour—before it must be renewed or terminated.

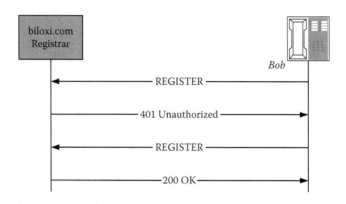

Figure 11.4   Registration-signaling flow.

How does Bob's UA obtain the URI of the registrar? It could be that the name is preprovisioned, or instead that Bob's UA locates a proxy server that has this information and is configured so that it always forwards REGISTER requests appropriately. We make no attempt to exhaust all possible scenarios.

Continuing through the signaling flow of Figure 11.4, we see that the biloxi.com registrar issues an authentication challenge in response to the first REGISTER request. This is the 401 Unauthorized message, which instructs Bob's UA to furnish authentication credentials. In the initial REGISTER request, a tag appears in the From header but not the To header. When it constructs the 401 Unauthorized response, the registrar will add a tag to the To header. (This treatment of tags is the same as in Section 11.4.) The UA provides the necessary credentials in the second REGISTER request; the 200 OK response indicates that those credentials have been accepted.

The reader should keep in mind that the signaling interchange described in the current section must precede that of Figure 11.2. Referring back to Figure 11.2, when Alice's INVITE arrives, the biloxi.com server is able to forward that INVITE to Bob using the information that it obtained *during Bob's registration procedure*. Here we assume that the biloxi.com proxy server and registrar can talk to one another. The proxy server and registrar could even be the same network element. In IMS, the Serving-CSCF plays both roles, but for SIP in general, this is neither discouraged nor required. In the case of the forking scenario of Section 11.3, Bob's wireless phone would also have registered using the same AoR as his desk phone.

Before moving on, we return briefly to the comparison between SIP and HTTP. HTTP lacks a registration functionality. This is consistent with what we have seen already—namely, the fact that HTTP URIs always lack user parts. In the contrasting case of SIP, SIP software registers on behalf of end users.

11.9 Feature Tags

To motivate the discussion in this section, let us return to a scenario in which Bob has a desk phone and a wireless phone. In addition, let us suppose that the desk phone supports interactive sessions with simultaneous voice and video, whereas the wireless phone lacks interactive video capability. Alice wants to initiate a multimedia (i.e., voice + video) session with Bob. How should this be handled?

To handle this situation in an intelligent way, the network needs to know that Bob's desk phone has the necessary capabilities, whereas his wireless phone does not. More generally, there are many scenarios in which a device needs a way to tell the network what it can and cannot do.

SIP's solution to this problem is to use *feature tags*, which are defined in [13]. A commonly used tag is the "methods" feature tag which, when appended to a Contact header, says which types of SIP requests a UA supports. Although we do not cover Contact headers or feature tags in detail, here is an example:

```
Contact:<sip:bob@biloxi.com>;methods="INVITE,BYE,OPTIONS,
ACK,CANCEL"
```

See Section 11.11 for a careful treatment of SIP request types.

11.10 Reliably Transmitting SIP Messages

In the signaling flows of Sections 11.1 and 11.2, recall that Bob's UA sends a 200 OK in response to Alice's INVITE. Upon receipt of that message, Alice's UA knows that the INVITE has reached Bob's phone. (We purposely skip over the 180 Ringing message in this discussion, as there are many INVITE flows in which it would not appear.) We say that the INVITE is transmitted *reliably*. In all normal cases, Alice's UA can expect a response, and will know something is wrong if it does not receive one.

Similarly, Bob's 200 OK is ACKed by Alice's UA. When the ACK reaches Bob's UA, it knows that the 200 OK was received by the far end. Thus, the 200 OK is sent reliably as well. When the original request is an INVITE, this is always the case. For non-INVITE requests, the responses are not sent reliably. Section 11.11 gives a full list of SIP requests.

We now turn our attention to the 180 Ringing response that appears in the example signaling flow. This indicates neither success nor failure; the ultimate fate of the session setup attempt is not yet determined at the time 180 Ringing is sent. 180 Ringing is an example of a *provisional* response—so named because one or more SIP messages are still forthcoming from Bob's UA, which is ultimately obligated to send a success response (200 OK is the most common of these) or a failure response. SIP responses are discussed in detail in Section 11.12.

Note well that the 180 Ringing response is *not* sent reliably—it is not ACKed.

11.11 SIP Requests

SIP is a request–response protocol. SIP requests are also called *methods*. The SIP requests that appear in the signaling flow of Figure 11.2 are INVITE, ACK, and BYE. The other SIP messages—180 Ringing, 200 OK—are responses. SIP responses are covered in Section 11.12. Except where noted otherwise, the methods that follow are defined in the baseline SIP specification [1].

11.11.1 ACK

ACK is classified as a SIP request. It is unusual among SIP requests in that there is no response to an ACK. ACK is the third part of the so-called *three-way hand-shake* for SIP INVITE transactions (i.e., INVITE/200 OK/ACK). Referring to the example given near the beginning of this chapter, Bob's SIP phone needs to be sure that Alice's PC has received the 200 OK. If it does not receive an ACK, Bob's SIP phone will suspect that something has gone wrong and will eventually trigger a resend of the 200 OK. Conversely, receiving an ACK tells Bob's SIP phone that it can go ahead and send (and also expect to receive) data.

All final INVITE responses are acknowledged with ACK. Classification of SIP responses into final and provisional categories is covered in Section 11.12.

ACK is not used at all outside of INVITE transactions. As noted in Section 11.11.5, INVITE is special: it is the only request for which responses are sent reliably. For instance, in our example signaling flows, there is no ACK following BYE/200 OK.

11.11.2 BYE

BYE is used to tear down a SIP session.

11.11.3 CANCEL

If a UA wants to cancel a request that it previously issued, it can do so by sending a CANCEL request. What happens if a UA sends a CANCEL request after the far end sends a final response but before the UA receives that final response? In this case, the CANCEL request has no effect. Final responses are covered in Section 11.12.

Stateful proxies process CANCEL requests differently from stateless proxies. Stateless proxies simply forward whatever responses they receive from "downstream" elements. On the other hand, stateful proxies respond directly to CANCEL requests rather than waiting to receive responses from downstream elements.

CANCEL is specifically intended for cancellation of INVITE requests. The base SIP specification [1] discourages but does not prohibit use of CANCEL for other SIP requests. CANCEL requests can be issued by SIP proxies as well as UAs.

11.11.4 INFO

The INFO request is intended to carry session-related information in cases where such information is generated after the session initiation signaling exchange is complete. To the best of our knowledge, INFO is rarely used but is included here for completeness. The INFO method is defined in [4].

11.11.5 INVITE

INVITE is used to set up a SIP session. Session parameters are negotiated during an INVITE transaction. These session parameters cover the type of session that is desired (e.g., voice, video, discrete messaging), the codecs that will be used, and bearer-plane addressing details. In the Offer/Answer Model, the initial offer is often carried in an INVITE. The Offer/Answer Model is covered in Chapter 12.

Why, among all SIP methods, is INVITE the only one for which final responses are ACKed? Referring back to the example of Figure 11.2, we note the following: once the session setup is complete, Alice and Bob are both free to speak. Before sending voice packets to Alice's PC, Bob's phone would like to know that his 200 OK has been received at the far end. (Otherwise, Alice's PC is not expecting to receive voice packets from Bob.) Once Bob's phone receives the ACK, it has this assurance. The fact that INVITE is tied to occurrences that are happening in real time is the reason for ACKing responses.

11.11.6 MESSAGE

The MESSAGE request is used to transfer Instant Messages. The body of a SIP MESSAGE is typically a text message, with or without an attachment such as a picture. The maximum size for a SIP MESSAGE—or, for that matter, any SIP request—is 1300 bytes. (Strictly speaking, this is a hard limit for some, but not all, transport layers. The 1300-byte limit ends up being a de facto limit across the board, however; otherwise, SIP networks with different transport layers would suffer from interoperability issues.) Assuming that 300 bytes is the average length of a SIP header, about 1 kilobyte remains for the message body. Here is an example SIP MESSAGE:

```
MESSAGE sip:bob@biloxi.com SIP/2.0
Via: SIP/2.0/UDP pc.att.net;branch=qK19035809
Max-Forwards: 70
To: Bob <sip:bob@biloxi.com>
From: Alice <sip:alice@att.net>;tag=dF26k3092
  .

  .

  .

Subject: Victuals
Content-Type: text/plain
Content-Length: 29

Let's get together for lunch.
```

In SIP MESSAGEs, the subject header is optional. As noted earlier, SIP message bodies are separated from headers by two consecutive CRLF sequences. That is why the blank line appears in the MESSAGE above. MESSAGE differs from other SIP methods in that the message body is intended to be displayed to an end user. In most cases involving other methods, the bulk of a SIP MESSAGE makes sense only to SIP software and is never directly seen by end users.

The MESSAGE method is defined in [5], which says that any response to a SIP MESSAGE must have empty body. The reasoning behind this is as follows. The authors of [5] wanted to make sure that the MESSAGE method would not be used to implement remote procedure calls. A remote procedure call is a request by a software process running on one machine to invoke a software process on another machine, which returns information regarding the outcome to the invoker. There are well-established means of performing remote procedure calls that are better suited to that task. The upshot is the aforementioned requirement that MESSAGE responses have empty bodies (thereby ensuring that outcomes could not be communicated to remote invokers). This requirement leads to inefficiencies in scenarios such as SMS messaging for Dual Mode Service. In Dual Mode Service, handsets are able to communicate with "the outside world" via cellular towers or, alternatively, through WiFi access points. If a Dual Mode handset needs to send or receive an SMS message when in WiFi connectivity, that need can be met by encapsulating the SMS message in a SIP MESSAGE. A SIP response (e.g., 200 OK) indicates the outcome in the SIP domain, but says nothing about success or failure at the SMS layer. To carry the SMS-layer success/failure information, a separate SIP MESSAGE must be generated.

If the recipient of a MESSAGE request wishes to send a reply (recall that the intended "consumer" of the MESSAGE body is an end user), then the recipient originates a new MESSAGE request. So there is a distinction between a SIP-layer response and an Instant Message reply from one end user to another.

11.11.7 OPTIONS

By sending an OPTIONS request, a SIP UA can ask another UA: "what do you support?" For instance, the response to an OPTIONS request usually indicates which SIP methods are supported by the recipient of the request; for some SIP methods, support is not mandatory. An OPTIONS response will also normally include information about acceptable formats for SIP message bodies. In the latter case, it is important to note that SIP itself does not care about formats of message bodies. SIP just provides a means for SIP entities to communicate about their capabilities and preferences in this regard.

A SIP UA can also send an OPTIONS request to a proxy server, but the reverse is not true; proxy servers do not issue OPTIONS requests.

Use of OPTIONS does not tell a SIP UA anything that it could not find out by initiating a signaling flow with INVITE or some other SIP method. In many cases, however, it may be preferable to find out "in advance" whether the requirements for a given service will be supported, rather than to attempt a service invocation and have it fail. As an illustration, perhaps the icon for invoking our hypothetical service should be "grayed out" whenever it can be determined in advance that service invocation attempts are doomed to fail.

If two SIP UAs want to use this method to gain knowledge of each other's capabilities, then OPTIONS requests must be sent in both directions. That is because the originator of an OPTIONS request is not declaring its own capability set, but instead, is only asking about the capabilities of the intended recipient.

SIP OPTIONS is not the only way to ascertain information about other UAs' capabilities. In Chapter 16, we will look at a presence framework that addresses such issues.

11.11.8 NOTIFY

The NOTIFY method is used to inform a SIP UA that a specified event has occurred. The NOTIFY method provides a very general notification capability in the context of SIP. Each time a NOTIFY request is sent, it must be tied to a so-called *event package*. Otherwise, the recipient UA will not be able to determine what sort of event has occurred. NOTIFY is usually discussed in the same breath as SUBSCRIBE, which is the most common means by which a UA can ask to be notified. The "classic" use (though not the only use) of SIP NOTIFY is in the context of presence and availability. Presence and availability is covered in detail in Chapter 17.

The NOTIFY method is defined in [6].

11.11.9 *Provisional Response ACKnowledgement (PRACK)*

To facilitate our discussion of the PRACK method, we need to update our signaling diagram with two things we have not seen in our previous examples: the

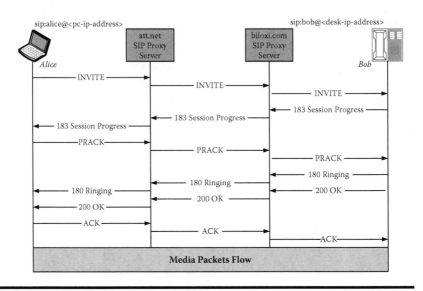

Figure 11.5   Example flow for the PRACK method.

183 Session Progress message (which, when used, is always a response to an INVITE) and PRACK itself. The updated flow appears in Figure 11.5.

Like 180 Ringing, the 183 Session Progress message is a provisional response. In the figure, Bob's desk phone sends a 183 Session Progress message to Alice's PC, which sends a PRACK request as a result. Like ACK, PRACK is classified as a request even though it is an acknowledgment of a response rather than an attempt to start a new signaling interchange. The 183 Session Progress message is useful if Bob's UA needs to provide a status update to Alice's PC prior to sending the 180 Ringing message. Additional information on 183 Session Progress is provided in Section 11.12.1.

In some cases, it is important that a provisional response be sent reliably. This is the purpose of the PRACK method. It is important to note that 183 Session Progress is not always PRACKed; it depends on the use case. The defining specification for PRACK 0 mentions use of PRACK to keep Bob's UA from unnecessarily retransmitting the 183 Session Progress message. In our experience, this is not common. There are other more realistic use cases characterized by the following:

■ Certain preconditions must be met before Bob's UA can proceed with a 180 Ringing message.
■ The PRACK message is used to communicate to Bob's UA that these preconditions are indeed met.

In the 3GPP™ IMS specifications (especially [9]), the classic use of PRACK is as follows: the service being invoked has quality of service (QoS) requirements. The network needs to verify that it can commit sufficient network resources so that Alice may experience an acceptable QoS. (Perhaps Alice is trying to invoke a video telephony application that has stringent resource requirements.) Once the network resources are secured, Alice's UA sends a PRACK to Bob's UA. Only then does Bob's desk phone alert him to the incoming call and, of course, send a 180 Ringing message.

In Figure 11.5, the 180 Ringing is not sent reliably (recall the discussion in Section 11.10). Although the standards do not prohibit PRACKing a 180 Ringing message, we have not seen any use cases that require it. As in previous examples, a BYE/200 OK exchange will take place when the session is complete. This has been omitted from Figure 11.5 because it adds nothing new.

11.11.10 PUBLISH

The PUBLISH method is employed whenever a SIP UA wants to publish status information. A UA can publish information about itself, or it can operate as a stand-in for another entity (e.g., a cellular phone or other device that is not SIP aware). As is the case with the NOTIFY and SUBSCRIBE methods, PUBLISH must always be used within the context of a well-defined event package. The PUBLISH method is defined in [10].

11.11.11 REFER

Let us embellish the example of Section 11.2 as follows: Alice calls Bob to ask him for specific information. Bob says that he is not the expert in these matters but that he normally goes to Ted with such questions. Bob could, of course, dictate Ted's SIP URI to Alice (suppose it is sip:ted@third _ party.com). Alice could disconnect the session with Bob, type in Ted's URI, and indicate that she wants to INVITE Ted to participate in a session with her. If Bob has requested that Alice let him know the outcome, Alice will then need to call Bob again or send a message to Bob by some other means indicating whether she was able to reach Ted.

All of this is rather clumsy. Alternatively, Bob could send Ted's URI to Alice, along with a request to call Ted. In terms of SIP User Agents, Bob's UA sends a REFER request to Alice's UA. Alice's UA then sends an INVITE to Ted, as indicated by the REFER request (perhaps the UI on Alice's PC asks for confirmation before launching the INVITE). Once the signaling interchange for the new INVITE is complete, Alice's UA sends a NOTIFY to Bob's UA indicating whether the session setup attempt succeeded. Notification of the outcome would, for example, allow Bob to REFER Alice to a different expert in the event that Ted is unavailable. The specification for the REFER method [11] indicates that notification of the outcome must be sent. After that specification was approved, new use cases arose in which the notification was superfluous. As a result, [11] has been updated by [12]. The latter document provides a means of "turning off" the notification mechanism.

11.11.12 REGISTER

We have already presented a registration signaling flow in Section 11.8. To summarize the scenario described there, we remind the reader that Bob must REGISTER before Alice can call him; the registration procedure is the means by which a binding between Bob's SIP AoR and one or more of his contact addresses is established. Note that multiple contact addresses can be bound to Bob's AoR in the same REGISTER transaction, although our example shows only one contact address. The fact that Bob is the called party—and therefore that the biloxi.com server must be able to find Bob—is the reason that we showed Bob's registration procedure instead of Alice's.

According to [1], it is not absolutely necessary for Alice to REGISTER prior to sending SIP INVITEs. This is not, however, practical from a carrier's point of view. The carrier (AT&T™, in this case) will insist that Alice's UA authenticate itself with its network before Alice is allowed to consume network resources. Recall that registration flows normally include an authentication step.

11.11.13 SUBSCRIBE

When a SIP UA wants to be notified regarding the status of another entity (aka, a "resource"), it does so by sending a SUBSCRIBE request. As is the case with NOTIFY and PUBLISH, SUBSCRIBE messages must always be tied to a well-defined event package. The most common use (but not the only use) for the SUBSCRIBE method is in the context of presence and availability, which is described in detail in Chapter 16. We present a simplified description here in order to put across the main idea behind the SUBSCRIBE method.

In the context of presence and availability, SUBSCRIBE requests are directed to a presence server. Initially, the presence server sends a NOTIFY to indicate the status of the subscribed-to resource at the time the SUBSCRIBE is received. (Note that the status of the subscribed-to resource is *not* sent in the 200 OK response to the SUBSCRIBE request. In this regard, the SUBSCRIBE/NOTIFY mechanism follows a similar philosophy to that of the SIP MESSAGE—see Section 11.11.6.) Subsequent updates, in the form of NOTIFY requests, are sent whenever the status of the subscribed-to resource changes.

The SUBSCRIBE method is defined in [6].

11.11.14 UPDATE

The UPDATE method can be used to modify parameters (e.g., codecs) associated with an INVITE that was previously sent. It is only used prior to completion of the signaling flow begun by the previously issued INVITE. The "calling party" (i.e., the UA that sent the INVITE) and the "called party" are both allowed to send UPDATE requests. Once session setup is complete, an endpoint wishing to alter session parameters would have to send a re-INVITE instead. (A re-INVITE is an INVITE request with its header fields set so they refer to an existing session.) The UPDATE method is defined in [8].

11.12 SIP Responses

Every SIP response can be classified as either a final or a provisional response, and is characterized by a 3-digit code. For all provisional responses, the first digit of that code is "1." The 3-digit code is the normative part (i.e., the portion of the SIP response that must comply with the standard). The text description that follows the 3-digit code can be altered. So, 200 Copacetic would, strictly speaking, be allowable, although we have never encountered this usage. Similarly, descriptive text comprising more than one word is allowable (as is the case for 183 Session Progress). Another curious fact is that the first line of a SIP response is arranged differently from that of a SIP request. By way of example, the first line of a 200 OK response looks like this:

SIP/2.0 200 OK

whereas the first line of a SIP request copied from an earlier section looks like this:

INVITE sip:bob@biloxi.com SIP/2.0

We are not certain of the motivation for placing SIP/2.0 at the beginning of the SIP response, but note that it may make the message parser's job easier (in the sense that the descriptive text ends just before the CRLF sequence and is therefore easy to find).

11.12.1 Provisional (1xx) Responses

The character of a provisional response is along the lines of "more signaling to come." The character of a final response is along the lines of "that is my final answer." Every SIP response is either provisional or final. Provisional responses to INVITEs are never ACKed. But, as we saw in Section 11.11.9, the PRACK method provides a means of acknowledging provisional responses in special cases.

Figure 11.6 shows three types of provisional responses, one of which was not presented in previous examples: 100 Trying. Two features of 100 Trying differentiate it from 183 Session Progress and 180 Ringing:

■ 100 Trying must not be PRACKed.
■ 100 Trying is sent hop-by-hop rather than end-to-end.

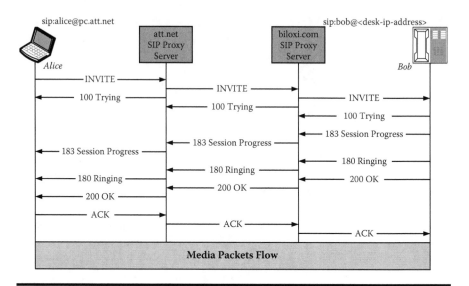

Figure 11.6   Signaling flow showing a variety of provisional responses.

To explain the second point, first note that time elapses as we travel downward in the diagram. So, for example, the 100 Trying that goes from Bob's UA to the biloxi.com server is sent after the 100 Trying that goes from the biloxi.com server to the att.net server. Similarly, the 100 Trying that goes to Alice's PC is the first one of the three. Each of these three messages is completely distinct from the other two, although all three pertain to the same INVITE transaction. In each case, the originator of the 100 Trying is informing the "previous hop" that it has received the INVITE and is following through.

This is in marked contrast to the 183 Session Progress and 180 Ringing messages. They are originated by Bob's UA and are forwarded by the two proxies in Alice's direction.

Before moving on, we note that 100 Trying is used very commonly. We omitted it from our examples earlier in this chapter not only for the sake of simplicity but also because it does not correspond to anything that the user perceives. In fact, it is used to adjust timers so that INVITE requests are not repeated unnecessarily. The 183 Session Progress response is not as commonly used as 100 Trying, but like 100 Trying, it does not correspond to anything that the end user would be aware of. This is different from 180 Ringing, which indicates that ringback should now be heard by the calling party.

For completeness, here is a list of standardized provisional responses. (We have already discussed those responses, which will be used extensively in this book.) Each of these responses is standardized in the baseline SIP specification [1].

- 100 Trying
- 180 Ringing

- 181 Call Is Being Forwarded
- 182 Queued
- 183 Session Progress

11.12.2 2xx Responses (Success)

When a request meets with a successful outcome, a 2xx response is sent (i.e., the first digit of the 3-digit response code is "2"). At this time, there are only two success responses:

- 200 OK, which is standardized in the baseline SIP specification [1].
- 202 Accepted, which is introduced in [6] (i.e., the same IETF RFC that defines the SUBSCRIBE and NOTIFY methods).

We have already encountered 200 OK. The 202 Accepted response is appropriate when a server has processed the request in question but must wait for something else to happen before it can complete the task at hand. As an example, perhaps an Instant Messaging server has received a SIP MESSAGE but has not yet been able to deliver that MESSAGE to its final destination.

11.12.3 3xx Responses (Redirection)

Here are the 3xx responses from the baseline SIP specification [1]:

- 300 Multiple Choices
- 301 Moved Permanently
- 302 Moved Temporarily
- 305 Use Proxy
- 380 Alternative Services

Here is an example illustrating the use of 302 Moved Temporarily. Suppose Bob happens to be on vacation when Alice tries to call him. Furthermore, suppose Bob has taken advantage of a call forwarding capability offered by his company. In this case, all of Bob's calls are forwarded to a colleague named Ted who works for the same company. This service can be implemented as shown in Figure 11.7. The biloxi.com server has access to Bob's forwarding settings. When it receives an INVITE directed to Bob's AoR, the biloxi.com server forms a 302 Moved Temporarily message containing Ted's AoR. Unfortunately, there was not enough room to include this information in the diagram; for completeness, we note that Ted's AoR is placed in the so-called *Contact* header of the 302 response.

Upon receiving the 302 response, the att.net server extracts Ted's AoR and constructs a new INVITE message with Ted's AoR in the To header. Upon noticing that Ted's AoR also belongs to the biloxi.com domain, the new INVITE is

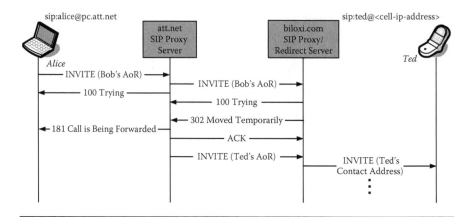

Figure 11.7   Illustration of 302 Response.

forwarded to the `biloxi.com` server. We have relabeled the `biloxi.com` server as a proxy/redirect server because it acts as a redirect server when processing the first `INVITE` and then as a proxy server. In this example, we assume that Ted is REGISTERed in the biloxi.com domain, and that his AoR is currently bound to a contact address identifying his cell phone. We emphasize that 302 is a final response, so the `INVITE` *branch* leading to Bob's AoR is dead once the 302 response is received. So the `INVITE` that is sent toward Ted is indeed a completely new message. All of this signaling takes time, so it might be appropriate to send a message to Alice telling her what is happening. This is the role of the `181 Call Is Being Forwarded` provisional response that appears in Figure 11.7.

The 3GPP IMS specification [9] (see Section 10.4.3, for example) also includes cases involving `302 Moved Temporarily`. There is a difference, however: in the 3GPP use case cited here, the IMS network is not able to handle the call. So the `302 Moved Temporarily` response travels all the way back to the end user's device, which can then attempt to place the call in the circuit-switched domain.

We saw a different call-forwarding scenario using the `REFER` method in Section 11.11.11. In that scenario, Bob needed to talk to Alice before deciding whom she could contact, as he needed to find a specialist to answer her particular question. The example of the current section differs in that the call forwarding is unconditional.

Before moving on, we note that the difference between `301 Moved Permanently` and `302 Moved Temporarily` is this: `301 Moved Permanently` indicates to the calling party that the contact information for the called party should be updated. A `302 Moved Temporarily` response carries no such implication. At the SIP layer, the signaling flow would be entirely similar if we had used `301 Moved Temporarily` instead.

11.12.4 4xx Responses (Request Failure)

Most of the examples in this book depict scenarios in which end users are successful in obtaining the services they want. We feel that such an approach is appropriate; the overarching goal of this book is, after all, to explore the kinds of innovate services that can be built atop the IMS architecture.

However, we would be remiss if we did not admit to the following variant on an old adage: "the devil is in the details *of the failure scenarios.*" To succeed in the marketplace, carriers must offer services that react gracefully when faced with nonideal circumstances. Thus, it should not be a surprise that there are many more failure responses (i.e., 4xx, 5xx, and 6xx responses) than provisional, success, and redirect responses. It is important to learn as much as possible about what has gone wrong and what might be done about it. With that preface, here is a catalog of 4xx response codes.

- 400 Bad Request
- 401 Unauthorized
- 402 Payment Required
- 403 Forbidden
- 404 Not Found
- 405 Method Not Allowed
- 406 Not Acceptable
- 407 Proxy Authentication Required
- 408 Request Timeout
- 410 Gone
- 413 Request Entity Too Large
- 414 Request-URI Too Large
- 415 Unsupported Media Type
- 416 Unsupported URI Scheme
- 420 Bad Extension
- 421 Extension Required
- 423 Interval Too Brief
- 480 Temporarily Unavailable
- 481 Call/Transaction Does Not Exist
- 482 Loop Detected
- 483 Too Many Hops
- 484 Address Incomplete
- 485 Ambiguous
- 486 Busy Here
- 487 Request Terminated
- 488 Not Acceptable Here

- 489 Bad Event
- 491 Request Pending
- 493 Undecipherable

The 489 Bad Event response is defined in [6]. All other responses are defined in the baseline SIP specification [1]. In Section 11.8, we saw that 401 Unauthorized is used to carry an authentication challenge in response to a REGISTER request. A 404 Not Found response will often be sent when the called party is not currently REGISTERed. This varies depending on the use case, however. For example, in the case of voice service, many implementations will autoforward to voice mail instead of issuing a 404 response. Other 4xx responses will be discussed only as needed throughout the book.

11.12.5 5xx Responses (Server Failure)

Here is a catalog of 5xx responses:

- 500 Server Internal Error
- 501 Not Implemented
- 502 Bad Gateway
- 503 Service Unavailable
- 504 Server Time-out
- 505 Version Not Supported
- 513 Message Too Large

All the responses listed here are defined in the baseline SIP specification [1].

11.12.6 6xx Responses (Global Failure)

Here is a catalog of 6xx responses.

- 600 Busy Everywhere
- 603 Decline
- 604 Does Not Exist Anywhere
- 606 Not Acceptable

All the responses listed in this section are defined in the baseline SIP specification [1].

Further Reading

RFC 3261, "SIP: Session Initiation Protocol," IETF, June 2002.

Alan B. Johnston, *SIP: Understanding the Session Initiation Protocol*, Second Edition, Artech House, 2004.

RFC 2616, "Hypertext Transfer Protocol—HTTP 1.1," IETF, June 1999.

RFC 2976, "The SIP INFO Method," IETF, October 2000.

RFC 3428, "Session Initiation Protocol (SIP) Extension for Instant Messaging," IETF, December 2002.

RFC 3265, "Session Initiation Protocol (SIP)-Specific Event Notification," IETF, June 2002.

RFC 3262, "Reliability of Provisional Response in the Session Initiation Protocol (SIP)," IETF, June 2002.

RFC 3311, "The Session Initiation Protocol UPDATE Method," IETF, September 2002.

TS 24.228, "Signaling Flows for the IP Multimedia Call Control Based on Session Initiation Protocol (SIP) and Session Description Protocol (SDP); Stage 3," 3GPP.

RFC 3903, "Session Initiation Protocol (SIP) Extension for Event State Publication," IETF, October 2004.

RFC 3515, "The Session Initiation Protocol (SIP) Refer Method," IETF, April 2003.

RFC 4488, "Suppression of Session Initiation Protocol (SIP) REFER Method Implicit Subscription," IETF, May 2006.

RFC 3840, "Indicating User Agent Capabilities in SIP," IETF, August 2004.

Chapter 12

Session Description Protocol

In Chapter 11, we looked at the SIP protocol for establishing and tearing down sessions. In this chapter, we will look at the use of the Session Description Protocol (SDP) in describing parameters carried within the SIP messages to describe multimedia sessions. SDP is based on RFC 4566 whose opening paragraph provides a concise purpose statement:

> When initiating multimedia teleconference, Voice-over-IP calls, streaming video, or other sessions, there is a requirement to convey media details, transport addresses, and other session description metadata to the participants. [5]

Thus, SDP will allow the originating and terminating parties[1] to describe their respective capabilities in establishing a session along with any desired media formats and port addresses.

As part of describing their capabilities, the parties will be able to negotiate or agree upon the session description values that will be used in their particular session. This is referred to as the Offer/Answer Model, which will be described in more detail in Section 12.2.

12.1 Field and Parameter Layout

The SDP session description appears in the SIP message after the following set of header fields, indicating an SDP session is to follow:

```
Content-Disposition: session
Content-Type: application/sdp
Content-Length: x (where x = length of the SDP body
                in bytes)
```

The session description must contain sufficient information to enable various parties to join the session, including any required resources. It is broken into three distinct sections in the following order:

- Session-level description—includes session identifiers and other session defaults for all media unless overridden by specific media section
- Timing description—describes start/stop/repeat times
- Media description—includes the type (e.g., video, audio) and format (e.g., H.263, MPEG4) of the media, transport protocol (e.g., RTP, UDP), address, and port details

Some SDP information may be present at either the session level or media level. If provided at the session level, it is global for all media sections. If provided at the media level, it will override any value that may have been provided at the session level.

12.1.1 Session-Level Description

The session-level description fields are shown in Table 12.1.

Each type must appear in the order shown in Table 12.1. Type values can be mandatory, optional, or conditional, depending upon certain occurring factors. Optional types may be omitted, but the type value that follows must still appear in order. Also note that Third Generation Partnership Project (3GPP™) has chosen not to implement all of the type values defined in RFC 4566 or has placed a different inclusion requirement on certain types. As SDP was intended to be embedded in multiple transport protocols, it would be expected that not all parts of the SDP protocol would be implemented in every transport protocol. Examples of transport protocols that would embed SDP parameters include SIP, Real-time Streaming Protocol (RTSP), Hypertext Transport Protocol (HTTP),

Table 12.1   SDP session-level description

Type	Description	RFC 4566	3GPP sending	3GPP receiving
v	Protocol version	m	m	m
o	Owner/creator and session identifier	m	m	m
s	Session name	m	m	m
i	Session information	o	o	m
u	URI of description	o	o	o
e	E-mail address	o	o	o
p	Phone number	o	o	o
c	Connection information	c	c	m
b	Bandwidth information	c	o	m
z	Time zone adjustment	o	o	n/a
k	Encryption key	o	n/a	n/a
a	Attribute line	o	m	m

Note: m—mandatory to include; o—optional to include; c—conditional; n/a—not applicable; RFC 4566—inclusion status of type described in RFC 4566; 3GPP—inclusion status within appropriate 3GPP/SIP message in Sending/Receiving mode.

Session Announcement Protocol (SAP), and e-mail using MIME extensions. A further explanation of the session level types follows:

Protocol Version (v): Current version of SDP. RFC 4566 defines version 0 so "v" is always set to "0."

    v=0

Owner/Creator (o): Provides the session originator, session identifier, and version number. It takes the form

    o=<username> <sess-id> <sess-version> <nettype>
       <addrtype> <unicast-address>

where
    <username> user's login on the originating host

`<sess-id>` used to form a globally unique identifier in conjunction with the other "o" values

`<sess-version>` a version number for this session description

`<nettype>` text string identifying the type of network (e.g., "IN" → Internet)

`<addrtype>` text string identifying IPv4 or IPv6 addressing (note other values may be designated in the future)

`<unicast-address>` address of the device from which the session was created

Session name (s=): Assigned by originator in text form

```
s=<session name>
```

Session information (i=): Provides textual information about the session

```
i=<session description>
```

URI (u): Pointer to any additional information about the session

```
u=<URI>
```

E-mail address (e=): Provides e-mail contact information for the person responsible in establishing the conference. More than one e-mail address may be provided.

```
e=<email-address>
```

Phone number (p=): Provides phone contact information for the person responsible in establishing the conference. More than one phone number may be provided.

```
p=<phone-number>
```

Connection Data (c=): Of the form

```
c=<nettype> <addrtype> <connection-address>
```

where

`<nettype>` and `<addrtype>` are the same as defined for "o="

`<connection-address>` indicates an IPv4 or an IPv6 address. If an IPv4 address is used, a time-to-live (TTL) value in the range 0–255 must be provided.

Bandwidth (b=): Indicates the proposed bandwidth for the session or media

```
b=<bwtype>:<bandwidth>
```

where

`<bwtype>` describes how the bandwidth should be interpreted. Defined values are "CT" for conference total and "AS" for application specific. Note that for "video" and "audio" media types that use RTP/RTCP, the "AS" descriptor is used.

`<bandwidth>` requested bandwidth in Kbps

Time Zones (z=): Used to schedule repeated sessions that span a change from daylight savings time to standard time and vice versa. The value is given in Network Time Protocol (NTP), which is the decimal time value in seconds since 1900.

```
z=<adjustment time> <offset> ...
```

Encryption Keys (k=): Provides a simple method for key exchange (note: the encryption key is not used in 3GPP networks)

```
k=<method>:<encryption key>
```

Attribute (a=): An attribute may be defined for both "session level" and "media level." A session-level attribute applies to the conference as a whole; the media-level attribute applies to an individual media. The attribute field may take one of two forms:
1. "Property" attribute of session of the form `a=<attribute>` such as `a=sendonly`.
2. "Value" attribute of the session of the form `a=<attribute>:<value>` such as `a=orient:landscape`

12.1.2 Time-Level Description

The time-level description fields are shown in Table 12.2.

The time-level types come immediately after the session-level types and follow the same ordering rules.

Timing (t=): Specifies the start and stop times for a session. The time is specified using NTP values.

```
t=<start-time> <stop-time>
```

Repeat times (r=): Used for sessions occurring at a regularly scheduled time.

```
r=<repeat interval> <active duration> <offsets from
   start-time>
```

Table 12.2 SDP time-level description

Type	Description	RFC 4566	3GPP sending	3GPP receiving
t	Time session is active	m	m	m
r	Repeat times	o	o	o

12.1.3 Media-Level Description

The media-level description fields are shown in Table 12.3.

The media-level types come immediately after the time-level types and follow the same ordering rules.

Media and transport (m=): Provides information about the media requested as well as transport. This field may be repeated if multiple media are requested.

```
m=<media> <port> <proto> <fmt>
```

where

<media> is the media type such as "audio," "video," "application," etc.
<port> is the port number where the media type can be received.
<proto> is the transport protocol such as "RTP/AVP," "UDP," etc.
<fmt> is the format list describing the media.

For example, if the <proto> field specifies RTP/AVP, then the <fmt> would contain the RTP payload type number. It can be noted that the dynamic fmt value used for a given codec need not be the same between the two UAs. But the tendency is to negotiate to the value used by the offerer (see Section 12.2). It is possible to have multiple <fmt> values as the UA may offer to the other UA a choice of formats. So, one example of the media parameter with two <fmt> values would look like

```
m = audio 2468 RTP/AVP 0 18
```

In the layout of the m= parameter, the <fmt> fields are presented as numeric values with their definition provided in RFC 3551. Those values from RFC 3551 are presented in the following two tables, with Table 12.4 showing the audio (A) payload definitions, and Table 12.5 showing video (V) payload definitions.

Table 12.3   Media-level description

Type	Description	RFC 4566	3GPP sending	3GPP receiving
m	Media name and transport address	m	o	m
i	Media title	o	c	c
c	Connection information	c	c	c
b	Bandwidth information	o	c	m
k	Encryption key	o	n/a	n/a
a	Media attribute line	o	o	m

Table 12.4   Audio-encoding payload definitions

Payload type	Encoding name	Media type	Clock rate (Hz)	Channels
0	PCMU	A	8,000	1
1	Reserved	A		
2	Reserved	A		
3	GSM	A	8,000	1
4	G723	A	8,000	1
5	DVI4	A	8,000	1
6	DVI4	A	16,000	1
7	LPC	A	8,000	1
8	PCMA	A	8,000	1
9	G722	A	8,000	1
10	L16	A	44,100	2
11	L16	A	44,100	1
12	QCELP	A	8,000	1
13	CN	A	8,000	1
14	MPA	A	90,000	Note 1
15	G728	A	8,000	1
16	DVI4	A	11,025	1
17	DVI4	A	22,050	1
18	G729	A	8,000	1
19	Reserved	A		
20	Unassigned	A		
21	Unassigned	A		
22	Unassigned	A		
23	Unassigned	A		
Dyn	G726-40	A	8,000	1
Dyn	G726-32	A	8,000	1
Dyn	G726-24	A	8,000	1

(continued)

Table 12.4   Audio-encoding payload definitions (continued)

Payload type	Encoding name	Media type	Clock rate (Hz)	Channels
Dyn	G726-16	A	8,000	1
Dyn	G729D	A	8,000	1
Dyn	G729E	A	8,000	1
Dyn	GSM-EFR	A	8,000	1
Dyn	L8	A	Var.	Var.
Dyn	RED	A	Note 1	
Dyn	VDVI	A	Var.	1

Note: Note 1—see RFC for additional definition.
"Dyn"—have no static payload type assigned; used only with dynamic payload types.
"Var."—can be a variable value.

Table 12.5   Video and combined encoding payload definitions

Payload type	Encoding name	Media type	Clock rate (Hz)
24	Unassigned	V	—
25	CelB	V	90,000
26	JPEG	V	90,000
27	Unassigned	V	—
28	nv	V	90,000
29	Unassigned	V	—
30	Unassigned	V	—
31	H261	V	90,000
32	MPV	V	90,000
33	MP2T	AV	90,000
34	H263	V	90,000
35–71	Unassigned	?	—
72–76	Reserved	N/A	N/A
77–95	Unassigned	?	—
96–127	Dynamic	?	—
Dyn	H263–1998	V	90,000

When the dynamic payload type is selected, additional information must be provided to describe it to the receiving end. This is accomplished by including an attribute line (a=) following the media line (m=). So, if you want to designate a 16-bit linear-encoded stereo audio sampled at 32 kHz, the SDP description might look like this:

```
m=audio 3456 RTP/AVP 96
a=rtpmap:96 L16/32000
```

Media title (i=): Similar in all aspects as the Session information field (including "i=" designation), except that it is used here for labeling media streams.

```
i=<media description>
```

Connection information (c=): Similar to the Connection information field in the Session-level. Not required if the Session-level Connection information is present and used as a global setting. A message can have this field in both the Session-level and the Media-level. If this occurs, the Media-level setting will override the Session-level setting.

```
c=<nettype> <addrtype> <connection-address>
```

For the "b," "k," and "a" fields, one can refer to their definitions under the Session-level description.

12.2 Offer/Answer Model

We indicated earlier in the chapter how one can use the SDP fields to inform the targeted receivers about the capabilities required for the session to be established. However, there were limitations in RFC 4566 when it came to the unicast or peer-to-peer model that is more prevalent in the telecommunications world. Thus, RFC 3264 was approved to address these limitations.[2]

In the Offer/Answer Model, the originating (owner) party wanting to create a new session, or who is already in a session and wants to modify the session to add on a new media component, will "offer" to the terminating party an SDP message[3] with a set of media streams, codecs, IP addresses, and ports that the originator (or offerer) would like to use to receive the new media component. The terminating party (or "answerer") will answer the request with an SDP message of its own with a matching media response for each media in the offer and indicating which media streams are accepted (along with the codecs, IP addresses, and ports) or not and optionally offering an alternative for rejected media.

Figure 12.1 shows several examples of an Offer/Answer exchange that may occur between the two parties. There are certain rules that must be adhered to during the Offer/Answer exchange process. These include

- The SDP message is exchanged reliably.
- New offers may not be made until any previous offer is answered.
- The initial offer must be in either the initial SIP INVITE message or in the first reliable response. *Note:* the SIP OPTIONS message may be used to determine the capabilities of the other party prior to sending the initial offer.
- If the offer is in the initial INVITE, then the answer must be in a reliable response.
- If the offer is in a reliable response, then the answer must be in the acknowledgment to that response.
- The number of media lines (m=) in the answer must equal the number of media lines in the offer.

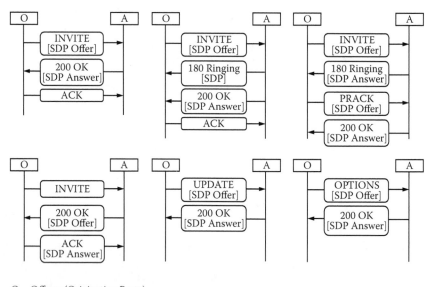

O – Offerer (Originating Party)
A – Answerer (Terminating Party)

Figure 12.1   Example of Offer/Answer exchanges.

Endnotes

1. These parties are referred to in the standards specification as user agents (UAs).
2. For example, a multicast session provides a global view to all participants in the session. For a unicast session, the two participants must provide information to each other in a two-way manner so that they can agree on what parameters to use between them. Additional examples are provided in [3].
3. The Offer/Answer Model, of course, assumes a higher-layer transport protocol (in this case, SIP) to convey the SDP message between the two parties.

Further Reading

3GPP TS 24.229, "IP Multimedia Call Control Protocol Based on Session Initiation Protocol (SIP) and Session Description Protocol (SDP); Stage 3."

RFC 3261, "SIP: Session Initiation Protocol," Rosenberg, J., Schulzrinne, H., Camarillo, G., Johnston, A., Peterson, J., Sparks, R., Handley, M., and Schooler, E., June 2002.

RFC 3264, "An Offer/Answer Model with the Session Description Protocol (SDP)," Rosenberg, J. and Schulzrinne, H., June 2002.

RFC 3551, "RTP Profile for Audio and Video Conferences with Minimal Control," Schulzrinne, H. and Casner, S., July 2003.

RFC 4566, "SDP: Session Description Protocol," Handley, M., Jacobson, V., and Perkins, C., July 2006.

Chapter 13

XML Configuration Access Protocol

XML Configuration Access Protocol (XCAP) is a means of remotely manipulating a document. The scenario is as follows: The document in question resides on a server, and an application running elsewhere (e.g., on a mobile handset) needs to read, update, or delete the document. Or, the application may need to create a new document on the server. The document in question cannot be in an arbitrary format; it has to be an eXtensible Markup Language (XML) document. Before discussing XCAP, we need some background on markup languages in general, and XML in particular.

13.1 What Is a Markup Language?

The easiest way to introduce the notion of a markup language is by means of an example. There are a number of markup languages that have seen widespread use, and XML is probably not the easiest place to start. Instead, we will look briefly at Hypertext Markup Language (HTML). Many Web pages are coded in HTML. HTML documents are always text documents. Here is a very simple HTML document:

```
<!DOCTYPE HTML PUBLIC "-//W3C//DTD HTML 4.01 Transitional//EN"
"http://www.w3.org/TR/html4/loose.dtd">
  <html>
  <head>
  </head>
  <body>
```

```
<strong>Smoke Signals and the Future of
   Telecommunications</strong>
<br>Class times for this exciting new course are as follows:
   <ul>
      <li>Wednesdays 6:30 - 8 PM</li>
      <li>Saturdays 9 - 10:30 AM</li>
   </ul>
</body>
</html>
```

When our sample HTML file is loaded by a Web browser, the display might look something like this:

Smoke Signals and the Future of Telecommunications
Class times for this exciting new course are as follows:
- Wednesdays 6:30–8 PM
- Saturdays 9–10:30 AM

The first thing to notice is that the portions of the HTML document that sit within the braces (i.e., "<>") do not appear in the display. The portions set off by braces are called *tags*, and the remainder of the document comprises the *content*. We say that tags are used to mark up the content; hence the term *markup language*.

The tag at the very beginning says that this is an HTML 4.01 document. For the moment, we will not worry about the additional information between the first "<" and the first ">." Everything between the <html> tag and the </html> tag (at the very end of the document) is HTML-encoded. For simplicity, the header here is empty (as there is nothing but a CRLF between the <header> and </header> tags). In real-world use cases, the header is rarely trivial.

Next we move into the body of the document, as indicated by the <body> tag. The text between the <strong> and </strong> tags is rendered larger than the other content in the sample display; it is also boldfaced. The details may vary from one Web browser to another—perhaps one browser defaults <strong> text to boldface, whereas another does not. The tag <ul> begins an unordered list; the tag </ul> ends that list. Within the list, there are two list items, each situated between a <li> begin tag and a </li> end tag.

Thus, HTML tags affect the manner in which the document content is displayed. For purposes of a forthcoming comparison with XML, it is important to note that the set of allowable HTML tags is fixed. Let's suppose, in the example from the beginning of this section, the tag <strong> was replaced by <max-ultra-strong> (and similarly for </strong>). A Web browser attempting to load the modified document would generate an error message because <max-ultra-strong> is not a valid tag in HTML, and there is no procedure for defining a new tag.

There is much more to HTML than we have shown here, but this suffices to introduce the notion of a markup language.

13.2 XML

We have seen that HTML is all about *displaying* data—the tags in an HTML document tell Web browsers how the content should look. XML, the markup language of primary interest in this chapter, is all about *describing* data. As is the case with HTML, XML documents are text documents.

Recall that XML stands for eXtensible Markup Language. Unlike HTML, XML is not specified with a fixed set of tags. That is the crucial feature that is extensible about this markup language: the author of an XML document can define tags that reflect the intended organization of the data at hand (i.e., the document content).

Here is a sample XML document:

```
<?xml version="1.0" encoding="UTF-8"?>
  <resource-lists xmlns="urn:ietf:params:xml:ns:resource-lists">
    <list name="pals">
      <entry uri="sip:joan@att.net">
        <display-name>Joan</display-name>
      </entry>
    </list>
  </resource-lists>
```

This XML document is the beginning of an example that we will weave throughout this chapter, and the concept of a resource list is crucial to the example. Resource lists can serve a variety of purposes, such as

- Buddy lists for instant messaging
- Contact lists in a more general "address book" setting
- Blacklists (i.e., lists of spammers that users do not want to hear from)
- Whitelists (e.g., to explicitly indicate that users are willing to receive promotions from selected merchants).

Of course, the first line of our sample document asserts that it complies with XML version 1.0. The following paragraphs cover important properties of XML documents.

XML documents are hierarchical. Like all XML documents, the example given in the previous section is hierarchical. The indentation style shown there reflects the layers of the hierarchy. However, the hierarchy is much easier to visualize diagrammatically. Figure 13.1 is a so-called tree diagram in which the node labeled "resource-lists" is the root. (The labeled circles are called *nodes*; computer scientists typically draw trees upside down.)

XML schema are used to specify XML document formats. Recall from Section 13.1 that the set of HTML tags is fixed. When it comes to XML, however, tags can be defined to suit a specific purpose. The tags in this example (e.g., <list> and

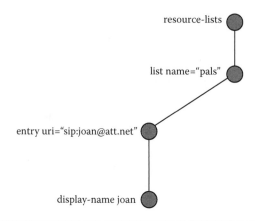

Figure 13.1   Diagrammatic view of sample XML document.

<entry>) are specified in a so-called XML schema. For completeness, the name of the schema is

```
urn:ietf:params:xml:schema:resource-lists
```

For convenience, we will simply refer to this as the resource-lists schema. Besides defining the allowable tags, the resource-lists schema specifies rules for document structure. For example, it says that a <resource-lists> item can contain an arbitrary number of lists (that is, <list> items). It also says that a <list> can contain an arbitrary number of entries.

The resource-lists schema has other functions, such as defining the document hierarchy. For example, a <resource-lists> item contains 0 or more <list> items. That is, <list> sits just below <resource-lists> in the hierarchy. Similarly, <entry> sits just below <list> in the hierarchy.

We have outlined some of the key document properties that XML schema are used to specify. We have chosen not to reproduce the resource-lists schema here, as it makes for tedious reading. It is important to note, however, that an XML schema is itself an XML document. Thus, the *same* piece of software that decomposes an XML document (such as the sample document near the beginning of this section) into its constituent components can also decompose schema. This is a clear asset from the software reuse point of view.

Specifying an XML schema is almost like specifying a new markup language—in our example, a resource-lists markup language. But, instead of building software from the ground up to deal with a new markup language, "generic" XML software can be used for fundamental capabilities such as parsing, verifying that a document is a well-formed XML document, and verifying that a document complies with a given schema.

XML and the manner in which the XML schema works is specified by the World Wide Web Consortium (aka the W3C). However, organizations other than W3C can specify new XML schema. This is by design, and it represents a significant advantage. Returning to our example, the resource-lists schema is defined by the IETF. In some cases (e.g., a schema is used only within a single company), a schema can be developed without the involvement of standards bodies or other industry fora.

XML namespaces are used to avoid conflicts between naming schemes. It would be naïve to think that the <resource-lists> schema would be the *only* schema with a legitimate use for a <list> tag. XML is widely used, and is applied to a broad variety of purposes. So, it is routinely the case that multiple developers (who probably are not even aware of one another) will define tags of the same name for differing uses. *XML namespaces* provide a means of formally defining naming schemes. Although the chance is little to none that <list> tags are unique to resource-lists documents, <list> tags *are* uniquely defined within the resource-lists namespace. Here is the second line from the sample XML document that appeared earlier in this section:

```
<resource-lists xmlns="urn:ietf:params:xml:ns:resource-lists">
```

The formal name of the aforementioned resource-lists namespace appears inside the quotes. (As the reader may have guessed, xmlns is an abbreviation for XML namespace.)

13.3 XCAP

Now we are ready to begin our discussion of XCAP in earnest, and the starting point for the discussion is this: XCAP is used to remotely edit XML documents. The first point to understand is that XCAP is not a standalone protocol. That is, there is no such thing as an XCAP packet. Rather, it is more accurate to regard XCAP as a tool for defining specialized usages of Hypertext Transfer Protocol (HTTP) and XML. The following paragraphs cover important properties of XCAP.

XCAP can only be used in the context of a well-defined Application Usage. An XCAP Application Usage specifies the points that client and server must agree upon beforehand in order for the tools at hand (i.e., HTTP and XML) to produce the desired results. We will continue to use resource-lists as an illustrative example. IETF RFC 4826 defines a resource-lists Application Usage. We have already discussed two key components of that Application Usage: its XML schema is

```
urn:ietf:params:xml:schema:resource-lists
```

and its default namespace is

```
urn:ietf:params:xml:ns:resource-lists
```

We will describe additional components of the resource-lists Application Usage as we proceed through this section.

XCAP uses HTTP URIs to identify XML documents and nodes within XML documents. Recall that XML documents are hierarchical. XCAP places the documents it manages within a larger hierarchy. Let us continue our resource-lists example as follows: suppose AT&T™ has a user, Ron Eng, whose SIP URI is sip:ron.eng@att.net. Ron wants to create a contact list that initially has only one contact—Joan. After Ron enters the necessary information, the client software on his device sends the following HTTP request to an HTTP server in the network:

```
PUT /resource-lists/users/sip:ron.eng@att.net/index HTTP/1.1
Host: xcap.att.net
...
Content-Type: application/resource-lists+xml
Content-Length: (...)

<?xml version="1.0" encoding="UTF-8"?>
<resource-lists xmlns="urn:ietf:params:xml:ns:resource-
    lists">
  <list name="pals">
    <entry uri="sip:joan@att.net">
      <display-name>Joan</display-name>
    </entry>
  </list>
</resource-lists>
```

The body of the HTTP PUT message (i.e., the portion after the content-length line) is the example XML file from Section 13.2. The HTTP server, whose hostname is xcap.att.net, responds with

```
HTTP/1.1 201 Created
...
```

The hierarchy resulting from this exchange is shown in Figure 13.2; note that the hierarchy from Figure 13.1 is contained within the new figure (it is the portion alongside the curly braces).

Each XCAP Application Usage has to have a naming convention. The HTTP URI from the PUT request above is

```
/resource-lists/users/sip:ron.eng@att.net/index
```

(We have not encountered an HTTP PUT request prior to this section. But, as is clear from context, it is a write request.) In that URI, the *filename* is index. How does Ron's device client know that the filename should be index when it issues the PUT request? Moreover, how does it know that index sits under the nodes

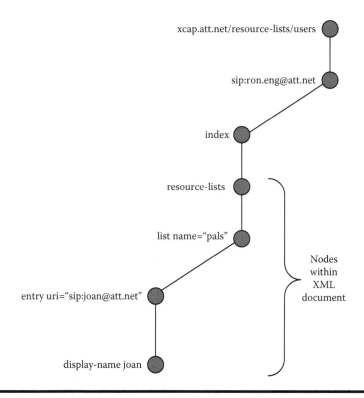

Figure 13.2 XML document of Figure 13.1 within XCAP hierarchy.

named resource-lists, users, and `sip:ron.eng@att.net` in the hierarchy? The client knows these things because they have been established ahead of time—this is the *naming scheme* for the resource-lists Application Usage. The naming scheme is different from the XML schema we discussed earlier. The difference is that the XML schema defines the hierarchy within an XML document, whereas the naming scheme says what the hierarchy should look like on the server—`xcap.att.net` in this case—but *outside* the XML document.

So far in this example, we have seen how a client can use XCAP to create an XML document on a server. XCAP also allows for clients to modify, read, and delete such documents. We will show how a document can be modified in this manner. We will not give read and delete examples, but the spirit is similar. Suppose Ron Eng wants to add a new contact, Bob, to the list that was created earlier. The client software on Ron's device issues a new HTTP PUT request with the following message body:

```
<entry uri="sip:bob@att.com">
  <display-name>Bob</display-name>
</entry>
```

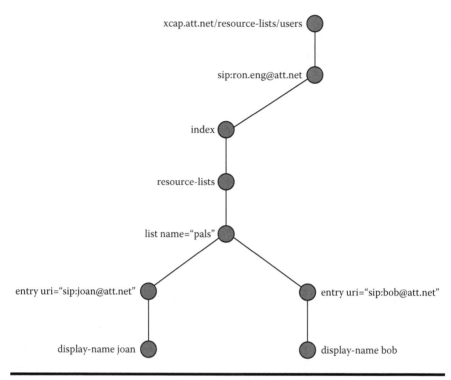

Figure 13.3   Hierarchy for Ron Eng after new contact is added.

Note that this message body consists of the new list entry, not the entire XML document. In real-world usage, we would encounter XML documents that are much larger than our example document. The ability to perform modifications without uploading the new document in its entirety is therefore very desirable. This is the power of XCAP.

We have omitted the details of the HTTP PUT request in the interest of simplicity. The HTTP URI in the PUT request is messy, but it is used to indicate something that is conceptually simple: it says where the new entry should be inserted. The upshot is that the hierarchy depicted in Figure 13.2 changes. Figure 13.3 shows the modified hierarchy.

13.4 Summary

XCAP relies on tools that are very widely deployed—particularly, HTTP servers and software for processing XML documents—and provides a streamlined means of accomplishing new tasks with those tools. This is consistent with the REpresentational State Transfer (REST) philosophy. REST is a popular approach to developing HTTP-based applications.

The example we worked through in this chapter is fairly complex. So, it may appear anything but streamlined to readers who are encountering this material for the first time. XCAP is streamlined in the following sense: there is a large community of application developers that is familiar with HTTP and XML. We have seen how a fairly sophisticated means of manipulating resource lists can be built using these familiar tools. It is worthwhile to note that HTTP proxies sitting between client and server do not have to understand anything at all about XCAP; so they do not have to be upgraded as XCAP Application Usages come along.

This chapter introduced XML and then moved on to XCAP. It is important to note that XML is used for many applications outside the context of XCAP.

Further Reading

"HTML 4.01 Specification," Raggett, D., Le Hors, A., and Jacobs, I., 24 December 1999, http://www.w3.org/TR/html4/references.html.

RFC 4825, "The Extensible Markup Language (XML) Configuration Access Protocol (XCAP)," Rosenberg, J., May 2007.

RFC 4826, "Extensible Markup Language (XML) formats for representing resource lists," Rosenberg, J., May 2007.

"HTTP—Hypertext Transfer Protocol," http://www.w3.org/Protocols/.

Representational State Transfer originated in Roy Fielding's Ph.D. dissertation, "Architectural Styles and the Design of Network-based Software Architectures."

Chapter 14

Message Session Relay Protocol

In Chapter 11, we saw how Session Initiation Protocol (SIP) is used to set up and tear down sessions. Chapter 12 covered the Offer/Answer Model. We summarize a few essentials from those chapters as follows:

- Three messages are present in a successful INVITE transaction: INVITE, 200 OK, and ACK. INVITE/200 OK/ACK is sometimes called the *three-way handshake*. More messages (e.g., 100 Trying, 180 Ringing) may be present.
- In the Offer/Answer Model, Session Description Protocol (SDP) session descriptions are exchanged during the 3-way handshake. In the most common approach, the *offer* is an SDP session description comprising the body of the INVITE request, and the *answer* is a session description comprising the body of the 200 OK response.

Our earlier coverage of SIP and SDP concentrated on sessions with so-called *continuous* media, for example, Voice-over-IP or video sessions. Once such a session is set up, Real-time Transport Protocol (RTP) is used to transfer, or *bear*, the voice or video media (stated differently, RTP is an example of a *bearer* protocol). However, neither SIP nor SDP is intrinsically tied to continuous media. So why not use SIP to set up sessions to exchange discrete messages? What about sharing files within the context of a SIP session? These questions serve as motivation for the material we present in this chapter.

To exchange discrete media such as messages and/or files, we first need to have a suitable bearer protocol. RTP is designed for continuous media and is therefore not appropriate for this type of task. Message Session Relay Protocol (MSRP) is purpose-built to fill this niche.

14.1 High-Level Description of MSRP

MSRP is a connection-oriented protocol for exchanging messages (and/or files) in the context of a session. MSRP does not place restrictions on message size or content type. There are two main MSRP standards: IETF RFC 4975 is the base specification, and IETF 4976 specifies relay extensions, which facilitate traversal of multiple networks (particularly when firewalls are in the message path).

MSRP functions within the context of a session. MSRP cannot set up sessions on its own, however. SIP is used to perform session setup and teardown. The MSRP specification does not rule out the use of some protocol other than SIP. However, to the best of our knowledge, nothing other than SIP is yet standardized for this purpose.

Figure 14.1 depicts setup and teardown of an MSRP session. When the session is set up, each user issues an MSRP SEND to transmit a message to the far end. The figure shows SIP and MSRP traffic going through the same messaging application server. (And strictly speaking, each handset is engaged in a separate MSRP session with the messaging application server. To the end users, however, this is indistinguishable from a single end-to-end session.) The following comments are in order here:

■ It is not necessary that the messaging application server be in the MSRP message path. SIP can be used to set up a so-called peer-to-peer MSRP session (i.e., a session in which the only two MSRP entities are User A and User B's devices). However, the configuration shown is likely to be a common one.

■ For simplicity, we have omitted signaling intermediaries (e.g., SIP proxies) from Figure 14.1. However, in most real-world scenarios, such intermediaries would be present in the SIP signaling path. In an IMS environment, IMS nodes such as the S-CSCF would, of course, sit between the end user and the messaging application server. However, SIP nodes such as the S-CSCF would be absent from the MSRP message path.

14.2 Key Features of MSRP

Unlike RTP, MSRP has its own URI scheme. It also has a chunking capability that facilitates fair sharing of connections. We will describe these MSRP features in the context of Figure 14.2.

In Chapter 12, we discussed SDP and the Offer/Answer Model, which is the vehicle for negotiating session parameters. For continuous media transmitted via RTP, endpoints' IP addresses are exchanged during the parameter negotiation process. MSRP does something a little more sophisticated: MSRP URIs are exchanged

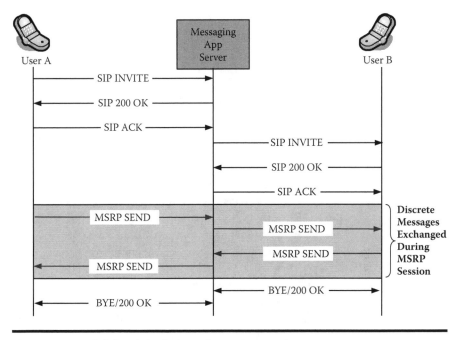

Figure 14.1   High-level depiction of an MSRP session.

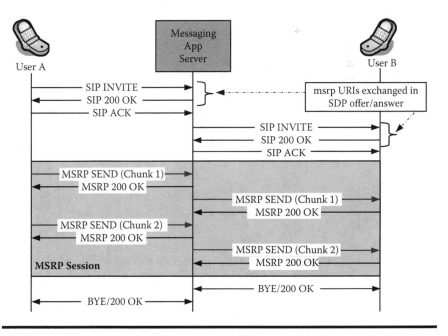

Figure 14.2   Detailed view of an MSRP session.

instead of "raw" IP addresses. The following MSRP URI is taken directly from RFC 4975:

```
msrp://biloxi.example.com:12763/kjhd37s2s20w2a;tcp
```

The name of the URI scheme (msrp in this case) is always followed by a colon character. Note that, unlike SIP URIs, there is no username or "@" sign. Many readers will recognize biloxi.example.com as a domain name. The "12763" following the second colon is a lower-layer identifier that is beyond our scope, as is "tcp" (which appears at the very end of the URI). That leaves a strange-looking character string following the last slash—kjhd37s2s20w2a in this example—which is called the *session-id*. For security purposes, this string is machine generated. RFC 4975 states that the session-id in an MSRP URI "must contain at least 80 bits of randomness." Provided that the SIP Offer/Answer exchange is protected from eavesdropping, the session-id will comprise a secret that is difficult for a malicious party to guess.

Once the MSRP session is set up, User A sends a message. As shown in Figure 14.2, User A's message is sent in two pieces, or *chunks*. (It only makes sense to do this if User A's message is large: a text message with a voluminous attachment, say. The number of chunks can be arbitrarily large; we have chosen 2 in this example for ease of exposition.) The MSRP software on User A's device is responsible for subdividing the outgoing message into chunks. Each MSRP SEND is answered by an MSRP 200 OK. As is the case with SIP, 200 OK is a success response. The MSRP software on User B's device is responsible for reassembling the chunks into a complete message.

The MSRP specification does not mandate that chunking be used, nor does it specify a preferred chunk size; these details are implementation specific. Given that there is no mandate to do so, why would implementers choose to use MSRP's chunking features at all? There are two main benefits:

- In cases—such as the example at hand—where messaging application servers are in the MSRP message path, chunking can improve end-to-end latency. Figure 14.2 explains the latency advantage schematically. The messaging application server can begin forwarding chunk 1 to session participants before it receives chunk 2.
- Chunking facilitates fair sharing of connections with other applications. Referring again to Figure 14.2, we see that User A's device has a connection to a wireless service provider's network. The messaging application in this example is using that connection to communicate with the messaging application server, and ultimately, with User B. Other applications—be they additional messaging sessions, other IMS applications, or something else altogether—may be trying to use the same connection simultaneously. Thus, another application may want to send packets between chunks 1 and 2. Chunking is especially beneficial in case a delay-sensitive application (e.g., a VoIP or video call) is coexisting with a messaging application that is sending messages with large attachments.

14.3 MSRP Relays

RFC 4976 introduces the notion of an MSRP relay and specifies authentication procedures. These procedures, which are separate from the authentication procedures described in Chapter 7, Section 7.2.2, are for clients to authenticate with MSRP relays at the MSRP layer. (Section 7.2.2 details IMS registration, which takes place in the SIP and Diameter domains and has a strong authentication component.)

MSRP relays do not appear in OMA and 3GPP™ specifications (e.g., OMA-TS-SIMPLE_IM-V1_0 and 3GPP TS 24.247). Evidently, the philosophy is that the IMS model for authentication and authorization is sufficient.

For completeness, we briefly discuss authentication between MSRP clients and MSRP relays. As stated in RFC 4976, two key reasons for defining MSRP relays are the following:

- To allow for a small number of inter-relay connections to carry messages for a large number of MSRP sessions. The end users participating in different sessions need not be the same.
- For messaging sessions that span multiple administrative domains, allow for each of the administrative domains involved to exert policy control.

Figure 14.3 presents an MSRP authentication flow. RFC 4976 defines the MSRP AUTH request. Assume that User A and MSRP Relay A are in one administrative domain and that User B and MSRP Relay B are in a different administrative domain. User A's device sends an AUTH request to relay A, which responds with an authentication challenge. This is quite similar to what we have already seen with SIP in Chapter 7, Section 7.2.2. User A's device sends another AUTH request, this time including its credentials, and receives a 200 OK in reply.

Next, User A's device authenticates with MSRP Relay B. It is important to note here that this authentication process flows through MSRP Relay A.

User B's device authenticates with MSRP Relay B and then with Relay A. As described earlier, the latter authentication flow goes through Relay B. The individual messages are suppressed here, but the flows are similar to those initiated by User A's device in every respect. The resulting symmetric arrangement allows for each administrative domain to enforce its own policies on subsequent MSRP messaging sessions.

User A's device can now issue a SIP INVITE. The SDP offer will include MSRP URIs indicating that both of the relays shown should be included in the MSRP message path. In the interest of simplicity, the details are omitted, as is the rest of the SIP message flow.

In the examples of Figure 14.1 and Figure 14.2, the messaging application server functions as both a SIP application server and an MSRP server. One key difference with the present example is that MSRP relays A and B need not be SIP application servers.

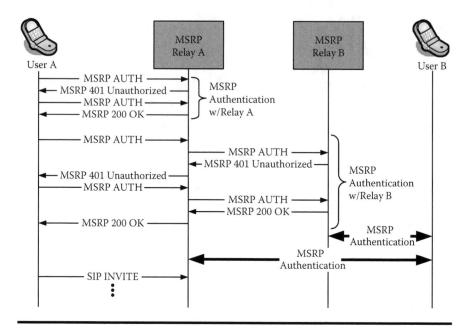

Figure 14.3   MSRP authentication flow with multiple relays.

14.4 File Transfer

MSRP can also be used to support file transfer applications. The Internet-Draft draft-ietf-mmusic-file-transfer-mech-11 specifies SDP extensions to characterize the files to be transferred. In the SDP Offer/Answer model, the offerer describes files that can be either

- Files the offerer wishes to send
- Files the offerer wishes to receive

In either case, the answerer can accept or reject the offer on a file-by-file basis. MSRP is then used to transfer the accepted files.

Further Reading

RFC 4975, "The Message Session Relay Protocol (MSRP)," Campbell, B., Mahy, R., and Jennings, C., September 2007.
RFC 4976, "Relay Extensions for the Message Session Relay Protocol (MSRP)," Jennings, C., Mahy, R., and Roach, A. B., September 2007.

Chapter 15

Diameter

Diameter is used to provide the Authentication, Authorization, and Accounting (AAA) functionality required by IMS. We can characterize the main components of AAA as follows:

- **Authentication** is the process of verifying that the user is who he or she claims to be. In the context of IMS, the client device (e.g., wireless handset) also verifies that it is registering to the right network—rather than a rogue network masquerading as such—in this step.
- **Authorization** is the mechanism for determining what a user is allowed to do. That is, authorization mechanisms are used to answer questions such as:
  - What services will be accessible to this user?
  - What information will be accessible to this user?
- **Accounting** is the mechanism for recording consumption of network resources by users. Among other things, accounting functionality is a prerequisite to usage-based billing. Accounting functionality is also important because it enables creation of audit trails.

It is important to be clear on the distinction between authentication and authorization. When a user tries to gain basic access to a secure network, the network issues an authentication challenge. (IMS authentication is covered in Chapter 7, Section 7.2.2.) That challenge, in effect, says: "Who are you? Show me your credentials." Once the credentials provided by the end user's device have been validated by the network, the authentication step has been completed. At this point, basic network access has been established. The network element responsible for issuing the authentication challenge is called a *network access server*. ("Network access server" is a generic term, as Diameter is also used in non-IMS contexts.) Credentials can take different forms; username and password will be familiar to most readers, but this is just one of many types of credentials.

Now that the question "who are you?" has been satisfactorily answered, the network needs to know: "What should you be allowed to do?" The latter is an authorization question. A simple parental control scenario serves as an illustrative example: a child has an account that allows basic Internet access from the family's personal computer (authentication), but that child is not allowed to access certain Internet sites because they are deemed to contain inappropriate content (authorization).

15.1 Motivation for Diameter

Diameter is not the only game in town, so to speak. Remote Authentication for Dial In User Service (RADIUS) is another AAA protocol. In fact, RADIUS is more widely deployed than Diameter as of this writing. To the best of our knowledge, Diameter is not an acronym; the name seems to have been chosen as a play on the word "radius."

Why was Diameter developed? The IP routers and network access servers of today are more complex than their predecessors. Moreover, these network elements must support scenarios that were not envisioned when RADIUS was developed. The classic example is that of authenticating with a network access server residing in one network through another network, and doing so in a secure fashion. To clarify: suppose a user wants to authenticate with a network access server managed by Internet Service Provider (ISP) A. So the network access server is sitting in ISP A's network. The complication is that our user can only access network A indirectly (through the network of ISP B, say). This scenario arises often: perhaps the user has traveled to a point outside of ISP A's realm of operation. How do we make sure that a malicious agent somewhere in network B does not "listen in" and steal the user's credentials?

It is not quite fair to say that RADIUS fails to address this scenario; RADIUS has evolved from its initial implementation. However, many extensions to "classic" RADIUS functionality are not standardized and therefore vary from one implementation to another. Moreover, RADIUS has some limitations that are very difficult to overcome with straightforward extensions. This motivated the development of Diameter.

15.2 Characteristics of Diameter

Diameter has the following key properties:

- Well-defined approach to reliability
- Extensibility and capability negotiation
- Explicit recognition of server hierarchy
- Mandatory security measures

15.2.1 Reliability

Unlike RADIUS, Diameter was designed from the beginning to provide a standardized, systematic approach to reliability. First, Diameter runs over reliable transport, whereas RADIUS was not initially designed to do so. "Reliable transport" means that the protocol layer running beneath the Diameter layer detects IP packet loss. Moreover, the lower-layer protocol initiates packet retransmission whenever packet loss is detected. One reason this is important is that Diameter messages may be large. Therefore, a Diameter message may be subdivided into multiple packets for transmission. The ability to retransmit starting with the first packet that was lost (rather than retransmitting an entire Diameter message) is a clear benefit.

Second, the Diameter specification includes well-defined failover behavior. Suppose that a network element (element A, say) tries to send Diameter packets to Diameter server B, but server B is down. In this case, lower-layer retransmissions will not succeed; network element A needs to identify and contact a backup Diameter server. "Well-defined failover behavior" means that the Diameter specification says exactly how this is to be done.

15.2.2 Extensibility and Capability Negotiations

Before discussing extensibility, we need a high-level understanding of Diameter message structures. Each Diameter message consists of a header followed by a set of attribute-value pairs (AVPs). Each AVP has a numeric code and a value. Here are two examples:

- Origin-Host AVP. The code for this is 264, and its value is the domain name of the node that originated this Diameter message. The Origin-Host AVP must be present in all Diameter messages.
- Disconnect-Cause AVP. Its code is 273, and is used when one of a pair of Diameter peers wants to disconnect from the other. This is a so-called enumerated AVP, meaning that there is a fixed set of allowed values. In this case, the number of allowable values is three:
  - 0 (which indicates REBOOTING)
  - 1 (which indicates BUSY, i.e., the disconnecting peer is experiencing overload conditions)
  - 2 (which indicates DO_NOT_WANT_TO_TALK_TO_YOU)

Diameter can be extended by creating new AVP values (e.g., extending the set of allowable values for an enumerated AVP), creating new AVPs, or creating new applications. Creating a new Diameter application is a bigger undertaking than either of the other two items, and normally includes creating new commands, AVPs, or AVP values.

In fact, a number of Diameter applications have been defined. We will touch on a few examples in Section 15.2.4. The idea is to leverage a general-purpose base protocol (as defined in RFC 3588) for a broad variety of purposes.

There is no requirement that every Diameter implementation support the same applications. For this reason, the design of Diameter includes a mechanism for capability negotiations. That is, Diameter provides a means for one Diameter node to ask another node which applications it supports. Details of the capability negotiation mechanism are beyond the scope of this book.

15.2.3 Server Hierarchy

As there are SIP servers of different stripes, including proxy and redirect servers, Diameter is designed to support a server hierarchy. (SIP proxy and redirect servers are described in Chapter 11, Section 11.6.)

Moreover, the usage of the terms *proxy* and *redirect* is consistent between SIP and Diameter. A Diameter proxy is a signaling intermediary: it forwards Diameter messages to another server on behalf of a Diameter client (and, of course, forwards messages in the opposite direction as well). Instead of forwarding messages from a client to another server, a Diameter redirect server sends a response telling the client to contact another server directly (and furnishes the information necessary for the client to do so).

Diameter's ability to accommodate a variety of server types allows for servers to be arranged in hierarchies, thereby facilitating large-scale deployments.

15.2.4 Security

RFC 3588 states that "Diameter MUST NOT be used without any security mechanism." Diameter messages must always be encrypted before they are transmitted. However, the details of the security mechanisms referenced in the Diameter specification are beyond our scope. We comment that the same security mechanisms can be used with RADIUS; the key difference is that RADIUS does not mandate that messages always be encrypted.

15.3 Diameter Base Protocol and Applications

15.3.1 Diameter Base Protocol

Most of the information we need about the base protocol has been covered in the preceding sections of this chapter. We recap that information here and add a few things to assist the reader with the following application descriptions. Figure 15.1 shows the interfaces that use the Diameter protocol in the IMS. These interfaces

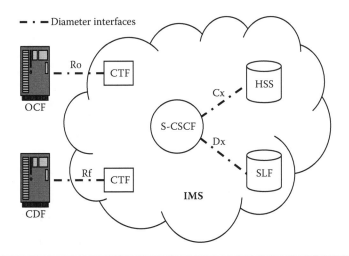

Figure 15.1   Diameter interfaces in the IMS.

were introduced in Chapters 4 and 8 and are highlighted here for completeness. They will be explored further in the following sections. Regarding the charging-related acronyms in the figure:

■ CDF stands for Charging Data Function.
■ CTF stands for Charging Trigger Function.
■ OCF stands for Online Charging Function.

The reader can refer back to Chapter 8 for details on these three functions.

Every Diameter message is a command, and every command is either a request or a response to a request. Every Diameter message consists of a header followed by a list of AVPs. The header indicates which command is being issued. AVPs are subdivided into mandatory AVPs and optional AVPs.

15.3.2 Diameter SIP Application

IMS makes use of the Diameter SIP Application (RFC 4740), as this application is the basis for the Cx and Dx interfaces (which connect S-CSCF to HSS and SLF, as shown in Figure 15.1). To give the reader a flavor of RFC 4740, we look at a command that is defined there: namely, the Push-Profile-Request (PPR) command. In the context of IMS, the PPR command is issued by the HSS whenever it needs to send an updated user profile to the S-CSCF. This would be necessary, for instance, when the user profile for a subscriber currently registered with IMS is modified (e.g., by provisioning activity).

Here is the format of the PPR command:

```
< Diameter Header: 288, REQ, PXY >
< Session-Id >
{ Auth-Application-Id }
{ Auth-Session-State }
{ Origin-Host }
{ Origin-Realm }
{ Destination-Realm }
{ User-Name }
* [ SIP-User-Data ]
[ SIP-Accounting-Information ]
[ Destination-Host ]
[ Authorization-Lifetime ]
[ Auth-Grace-Period ]
* [ Proxy-Info ]
* [ Route-Record ]
* [ AVP ]
```

Regarding the first line, which is the message header: 288 is the numeric code for the PPR command, REQ indicates that this is a request, and PXY indicates that proxy servers are allowed to forward this message. The remaining lines denote AVPs. The angled brackets "<>" surrounding the Session-Id AVP indicate that it must appear immediately after the header. Particulars of the Diameter Session-Id are beyond the scope of this book. The curly braces (i.e., "{}") surrounding subsequent AVPs indicate that those AVPs are mandatory, but that their order does not matter. The square braces (i.e., "[]") surrounding the remaining AVPs indicate that those AVPs are optional (and order does not matter). Lastly, the "*" preceding some of the AVPs indicates that they are allowed to appear more than once.

The User-Name AVP indicates which user's profile is to be updated. The profile information itself—for example, initial Filter Criteria (iFCs)—is contained in SIP-User-Data AVPs. Recall from Chapter 6 (particularly Section 6.1) that the S-CSCF uses iFCs to determine which application servers it should invoke, and under which circumstances these invocations should take place. In fact, this is the "frontline" authorization mechanism of IMS (the second "A" in AAA). Note that application servers can be used to implement additional authorization features.

IETF does not specify the format of SIP-User-Data AVPs. For IMS, that format is specified as part of 3GPP TS 29.228, which defines the Cx interface. More specifically, Third Generation Partnership Project (3GPP™) has published an XML schema—as an adjunct to 3GPP TS 29.228—defining the required format for "value" portions of SIP-User-Data AVPs.

We mention, but do not detail, two other Diameter commands defined in RFC 4740: namely, the Server-Assignment-Request (SAR) and Server-Assignment-Answer (SAA) commands. Those commands appear in Chapter 7, Figure 7.3, which depicts an IMS registration flow. The figure is accompanied by a careful description of authentication aspects of the flow.

In Figure 7.3, the S-CSCF issues the SAR messages, and the HSS sends an SAA message in response. We add here that the SAA message contains the iFCs for the user in question, as is the case with the PPR message detailed earlier. In fact, the S-CSCF usually obtains iFCs upon request via SAA messages from the HSS; this happens during every IMS registration. The PPR message is used for the less common occurrence of a profile change taking place while a user is registered with IMS.

15.3.3 Diameter Credit-Control Application

Up to now, we have said very little about accounting (the third "A" in AAA). Accounting functionality is crucial for a number of reasons. Among other things, accounting is necessary to support billing and charging.

The prepaid market represents an important customer base for many wireless operators. As described in Chapter 8, Section 8.2.2, the Online Charging System supports IMS services for prepaid customers. The Online Charging System interacts with the rest of the IMS domain via the Ro interface, which is a Diameter interface. To see the positioning of the Ro interface, the reader can refer to Chapter 8, Figures 8.5 and 8.6.

To round out the current chapter, we briefly describe the Diameter credit-control application, which is defined in RFC 4006. The Ro interface specification (3GPP TS 32.299) is essentially an extension of RFC 4006. RFC 4006 defines two (and only two) new commands: Credit-Control-Request (CCR) and Credit-Control-Answer (CCA).

RFC 4006 also defines a number of new AVPs. We will confine ourselves to the following two examples:

- Requested-Action AVP. This is an enumerated AVP whose allowable values are
  - 0 (indicating DIRECT_DEBITING)
  - 1 (indicating REFUND_ACCOUNT)
  - 2 (indicating CHECK_BALANCE)
  - 3 (indicating PRICE_ENQUIRY)
- Check-Balance-Result AVP. This is also an enumerated AVP. Its allowable values are
  - 0 (indicating ENOUGH_CREDIT)
  - 1 (indicating NO_CREDIT)

The Requested-Action and Check-Balance-Result AVPs can be used as follows. When a prepaid customer tries to invoke a service, the network needs to know whether the account balance is sufficient for this service invocation. A CCR message is sent to the Online Charging System. The command indicates the price of the service invocation; the Action-Requested AVP is set to indicate CHECK_BALANCE. The Online Charging System will send a CCA message with the

Check-Balance-Result AVP set to indicate whether there is sufficient money in the customer's account to pay for this service. The service invocation attempt will accordingly be granted or refused by the network.

Further Reading

RFC 3588, "Diameter Base Protocol," Calhoun, P., Loughney, J., Guttman, E., Zorn, G., and J. Arkko, September 2003.

RFC 4006, "Diameter Credit-Control Application," Hakala, H., Mattila, L., Koskinen, J.-P., Stura, M., and J. Loughney, August 2005.

RFC 4740, "Diameter Session Initiation Protocol (SIP) Application," Garcia-Martin, M., Belinchon, M., Pallares-Lopez, M., Canales-Valenzuela, C., and K. Tammi, November 2006.

3GPP TS 29.228, "Technical Specification Group Core Network and Terminals; IP Multimedia (IM) Subsystem Cx and Dx Interfaces; Signalling Flows and Message Contents."

3GPP TS 29.299, "Technical Specification Group Service and Systems Aspects; Telecommunications Management; Charging Management; Diameter Charging Applications."

IMS/OMA-BASED ENABLERS

Chapter 16

Presence

The concept of Presence became part of our everyday communication pattern with the rise in popularity of the various instant messaging (IM) communities due to the proliferation of the Internet and personal computers. Being able to set one's presence status to let your "buddies" know that you are available to chat is seen as a powerful communication tool. It was with this paradigm in mind that the designers within Third Generation Partnership Project (3GPP™) and Open Mobile Alliance (OMA) set out to extend the concept of presence from just a pure IM feature to a general-purpose feature that can be used across multiple devices as well as across multiple applications.

The term *Presence* has undergone an expansion in the scope of its definition to cover three main areas:

- Presence
- Availability
- Mood

The purist definition of Presence would only cover whether the subscriber is logged onto the network. We see from our earlier IM example that the user must sign into their IM service to be "present." In the case of a mobile user, the mobile device must be powered on and registered with its home network. However, we know that just being connected to the network does not mean that the user is willing and able to communicate, or will do so in a civilized manner.

This leads us to the two other aspects of Presence, namely, *availability* and *mood*. Availability is an indication of the user's willingness to engage in a particular communication form based on his or her current status. Here, a user may be going into a meeting or a movie theater (some place where they are not at liberty to have a verbal conversation). The user may wish to indicate that they are only available via a text message and not through any other means. Alternatively, the user may have an application that looks at their calendar and will provide an available/unavailable status automatically based upon the time and interruption status of a particular event. For this example, the application could have all phone calls sent directly to the user's voice mailbox automatically, independently of any user action.

Finally, Mood is an indication of your current emotional state and can serve as either an invitation to communicate or a warning to communicate carefully, if at all. Mood is another concept brought over from the IM world and generalized to be applicable for other communication methods.

There are several standards currently available for presence. Many of these standards, however, are application specific and thus do not fit the IMS model. This was recognized in OMA which has oversight for the OMA Instant Messaging and Presence Standard (IMPS) standard (formerly Wireless Village). The OMA IMPS is limited to supporting IM applications and presence information that cannot be shared across applications. Thus, this standard was capped at its version 1.3 specification. Recognizing this, the OMA Presence standard produced in the SIMPLE working group is designed for the IMS environment to be multiapplication in nature. This chapter will focus on the OMA Presence SIMPLE (SIP for Instant Messaging and Presence Leveraging Extensions) standard, along with the corresponding 3GPP and Internet Engineering Task Force (IETF) standards.[1]

16.1 Presence Architecture

The Presence architecture consists of three players with their own unique role: presentity, presence server, and the watcher (Figure 16.1). Each of these players will now be discussed in more detail.

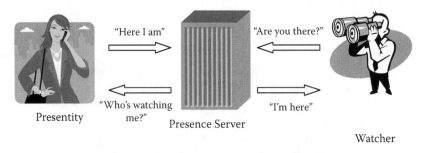

"Here I am" "Are you there?"

"Who's watching me?" "I'm here"

Presentity Presence Server Watcher

Figure 16.1 Presence players.

16.1.1 Presentity

The term *presentity* is a combination of the words *presence* and *entity*. Thus, a presentity is any unique entity that presence information can be obtained for and which would be stored in the Presence Server. A presence agent that can take the form of a presence user agent (PUA) or a Presence Network Agent (PNA) makes the presence information of the presentity available to the Presence Server.

The PUA is typically found in the IMS mobile terminal or end user's device and provides presence status information directly to the Presence Server. The PNA collects information about a presentity when the presentity's terminal device cannot provide its information to the Presence Server directly. An example of this would be a presentity with a 2G (non-IMS) device that cannot send its presence information directly to the Presence Server. The PNA would collect the presence information through other network elements. In one example, the PNA could query the HLR to determine if the presentity has registered (and thus is "present" on the network).[2] Figure 16.2 shows the defined network elements that the PNA can connect to along with the defined reference point for the interface.

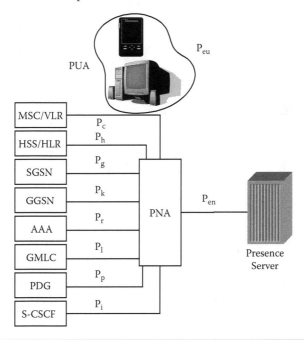

Figure 16.2   PUA and PNA relationship to the presence server.

16.1.2 Presence Server

The Presence Server acts as the central repository for the storing of the subscriber's presence information. It also acts as the distributor of the subscriber's (presentity) information to approved watchers. The term *Presence Server* is also a generic term that can refer to all the components that make up a Presence Server product. The OMA architecture defines four functional components that make up a Presence Server product (Figure 16.3).

- Presence Server: Collects and distributes presence information on the presentity.
- Presence XML Document Management Server (XDMS): Contains list information in an XML document that is specific to the presence service. One example can be the list of authorized watchers for the presentity's presence information.
- Resource List Server (RLS): Manages the various presence list subscriptions for a watcher. In this manner, a watcher subscribing to the presence information of multiple presentities needs to send only a single transaction request for a particular list as opposed to a single transaction request for each presentity in that list.
- Resource List Server XDMS: Contains the actual list data used by the RLS in an XML document. This allows for a change in a single subscription to be propagated to multiple documents stored in the RLS XDMS.

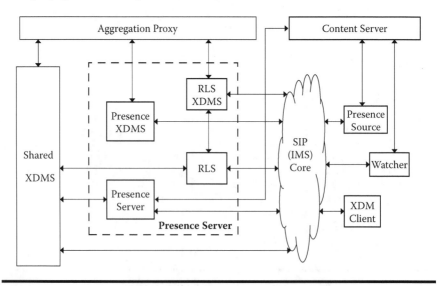

Figure 16.3   Presence reference architecture.

In addition to the four functional components within a Presence Server product, OMA had defined four additional functional components as part of the Presence reference architecture (Figure 16.3):

∎ Presence source would typically be the PNA or PUA as discussed previously.
∎ Content server is a functional entity that manages Multipurpose Internet Mail Extensions (MIME) objects.
∎ Watcher is any entity that makes a requests of the presence server for presence information about a presentity or watcher information about a watcher (see Section 16.1.3 for more details).
∎ XDM client manages the eXtensible Markup Language (XML) documents associated with Presence (e.g., authorization lists).

The shared XDMS and aggregation proxy interact closely with the Presence Server and will be discussed in more detail in Chapter 17.

16.1.3 Watcher

The watcher is the subscribing party who is interested in knowing the presence status of one or more presentities. The watcher can subscribe with the Presence Server to request presence information on a particular presentity or to be notified when a change in the presence status of that particular presentity occurs.

Of course, for privacy and security reasons, the watcher is not allowed access to the presence information of the presentity unless the presentity has granted permission to that watcher. The presentity is informed of a watcher trying to obtain its presence information by itself subscribing to the Presence Server for any change in a watcher's information state (e.g., new watcher requesting information). Here, the presentity subscribes to the Presence Server using a watcher-information (winfo) package that requests new watcher information. When the presentity receives the notification of the request, they will have the option to authorize or deny that particular watcher. This authorization policy is sent to the Presence XDMS, which will store these settings and in turn will forward/trigger the Presence Server to apply the updated policy.

16.2 Protocols for Presence

We saw in Chapters 11 and 12 how the SIP messages carrying the SDP parameters were used to set up and negotiate a session. In a similar manner for Presence, SIP messages carrying the Presence Information Data Format (PIDF) parameters are used to convey presence information.

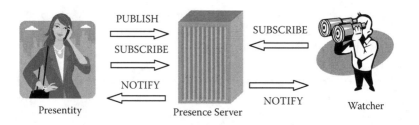

Figure 16.4 SIP messages for Presence.

Presence makes use of three SIP methods (messages) as shown in Figure 16.4.

PUBLISH: The presentity sends to the Presence Server their current presence information (state), which may include their availability and mood status.

SUBSCRIBE: A watcher wishing to obtain the presence status of a presentity will subscribe for this service from the presence service. A presentity would also subscribe to the Presence Server to be notified of who is watching them (i.e., winfo package).

NOTIFY: Used to inform a subscriber of a change in state or event. This is an asynchronous event.

The presentity will typically publish (by sending a PUBLISH message) their presence status upon a change in event that may or may not be initiated by the subscriber (i.e., presentity). An automatic publish may occur when the subscriber powers up or down their mobile device, changes location, or changes networks. The changing networks can occur when the subscriber moves between different access networks such as 2G, 3G, Wi-Fi, VoIP, or a private (enterprise) network. It can also be triggered as the subscriber roams onto another carrier's network.

A manual publish will occur when the subscriber takes a physical action (e.g., pushes a soft key, provides login information, etc.). This can occur when the subscriber sets his or her availability or mood status or does a manual push of his or her presence state. The different values (or settings) for the subscriber to set are based on the PIDF specification (RFC 3863) and the Rich Presence Information Data format for presence (RPID) specification (RFC 4480), which defines additional service attributes. These service attributes can refer to a person, a device, or a particular service.

The OMA Presence draws upon several RFCs to provide these service attributes. Among the referenced RFCs are presence data model [10], [14], location types [15], and geographical objects [13]. Many of these service attributes have fields that are populated by the subscriber to describe their particular situation or experience. As such, these attributes are optional within the SIP messages. The attributes (or elements) that have been defined related to a subscriber's availability or mood that can be incorporated into a service include the following:

- Activities: Describes what the person is currently doing, such as an appointment, having dinner, in a meeting, or on holiday. This information could also be derived from a calendar application.
- Class: The presentity can identify the class of a service, device, or person.
- Device ID: References a device that provides a particular service. Useful when the presentity moves between different devices (e.g., mobile phone or VoIP phone) or performs SIM swapping between devices.
- Mood: Describes the mood of the presentity. Over 60 different moods have been defined [10] and range from afraid, to worried, to invincible.
- Place-is: Describes the atmosphere of the current location of the presentity. This information can be useful to the watcher in order to know how best to communicate. These descriptions have been subdivided into three categories where one, two, or all three subcategories can be provided.

  1. Audio: An example is "noisy," which means it is difficult to carry on a verbal conversation.
  2. Video: An example is "too bright," which indicates the presentity could send the watcher a video share call because the lighting is good.
  3. Text: An example is "inappropriate," as the user could be in a house of worship where it would be inappropriate to text messages.

- Place-type: Describes the type of place the presentity is currently located. This can be a name of an establishment such as a restaurant or a generic phrase such as "church."
- Privacy elements: Another way to describe current environment of the presentity (note: this does not refer to who can see their presence information). This is intended to let the watcher know if other people are around who could inappropriately overhear a conversation or see a video communication.
- Relationship: Identifies the type of relationship the presentity has with a contact such as family, friends, or associates. A URI is designated for the contact as an alternative for the presentity.
- Service class: Designates a type of communication exchange service such as electronic, postal, courier, freight, in-person, or unknown. Additional data elements will define a physical service delivery address. An electronic service will be defined by a URI.
- Sphere: Designates the current mode the presentity is in, such as their home sphere or work sphere or their participation in some organizational activity such as "Boy Scouts." This will allow applications to apply certain rules to a particular group of contacts such as coworkers in a work sphere or scouts' parents in a Boy Scouts sphere.
- Status-icon: Provides a URI pointer to the icon (image) representing the current status of the presentity for presentation on a watcher's graphical user interface (GUI).

- Time offset: Provides the presentity's current local time in relationship to Universal Time Coordinates (UTC).
- User-input: Records the user-input or usage state based on input by the presentity.

16.3 Privacy

The world of online communities has opened up a new area of users revealing information about themselves. However, although users may want to reveal personal information about themselves to enable a particular service, users also do not want that information revealed to just anyone, but instead would like to have control over who sees their information.

We saw in the preceding section how users (the presentity) can subscribe to know who wants to see their information. This gives the user the opportunity to know who wants to see their information. When a request from a watcher is sent to the presentity, the presentity has one of four options in responding:

- Allow
- Confirm
- Block
- Polite blocking

Here, *Allow* permits the watcher to see information permitted by rules defined by the presentity and usually established a priori. *Confirm* places the request in a pending state (e.g., for further action). *Block*, as its name implies, rejects any requests for information on the presentity. *Polite blocking* is a way to indicate to the watcher that his or her request for the presentity's information has been approved; however, no actual presence information will be sent to this particular watcher.[3]

16.4 Traffic Optimization

Although the delivery of presence information is defined pretty straightforward, a major engineering problem can be created for the network operator as the popularity of the service grows. It is recognized that the messaging traffic for presence can begin to scale exponentially as the number of presence users grows linearly. As a quick example, if we have a closed network (hypothetical example) of ten (10) presence users who are in each other's presence-enabled address book (buddy list), then for each of them to get the presence status information of each contact in their

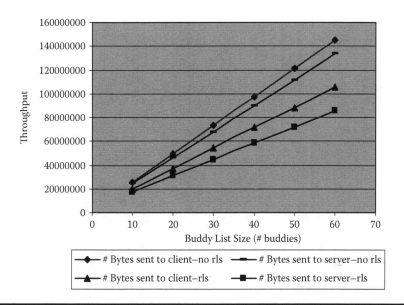

Figure 16.5   Impact of buddy list size on throughput. (Source: Nokia-Siemens Networks.)

address book, each user must make nine requests and in turn would receive nine responses from the Presence Server for $N \times [2 \times (N - 1)] \times 2$ (or 360)[4] messages sent within the network for a single buddy list. If we allow a customer's buddy list to grow in size, we see there is a linear growth in the throughput requirements (Figure 16.5). Savings in message throughput can be realized if an operator implements the RLS function described earlier in the chapter.

Although the preceding example is contrived, it should be clear that a large number of messages will be generated impacting many portions of the network as presence-based services take off in popularity (Figure 16.6). This will have a major impact on network operators, whose packet switching elements and transport backbone must scale to meet this new traffic demand. A further burden will be placed on wireless operators as the air (radio) interface is a limited bandwidth resource to be shared among many users.

With this in mind, mechanisms need to be defined to limit both the number of presence-related messages as well as the size of these messages, to prevent overburdening of the network. One can control the number of messages by limiting when a message can be sent or by combining multiple messages into a single message (thus reducing the total byte throughput by efficiency savings in overhead fields). Limiting the number of messages can be provided by the following:

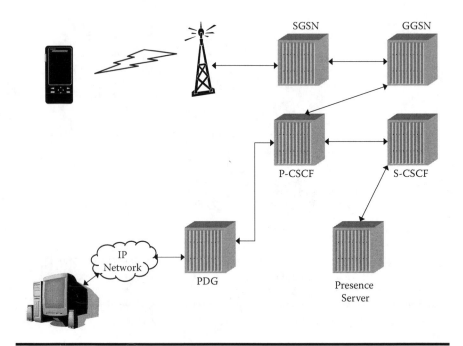

Figure 16.6   Presence message impact through operator's network elements.

- Configuration of time between PUBLISH messages: The Presence Server can specify to the PUA or PNA the time interval permitted between updating (publishing) their current presence status (Figure 16.7).
- Event throttling: The Presence Server restricts the time between NOTIFY messages sent to a watcher. A minimum time period is specified, usually as a configurable setting before a new notification message is sent (Figure 16.8).
- Event notification filtering: The watcher can specify the content and the triggers for when a notification occurs (Figure 16.9).
- Resource list server (RLS): The watcher can request the status of an entire group of presentities with a single request referencing that particular group.
- Pull/push notification: The application can require the user to pull status information only when the user needs the information (user/watcher must take a physical action to retrieve the presence status for his or her list of presentities). Likewise, the Presence Server could push the information to the watcher only when a change occurs with the presentity and not allow the watcher to constantly ping it for update information (Figure 16.7).

The size of the presence messages can also be limited by taking an approach whereby only a change in a setting needs to be conveyed as a record of previous settings is maintained. For example, when a presence-enabled network address book

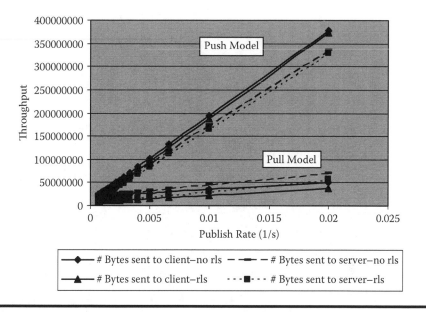

Figure 16.7   Impact of publish rate on throughput push versus pull model. (Source: Nokia-Siemens Networks.)

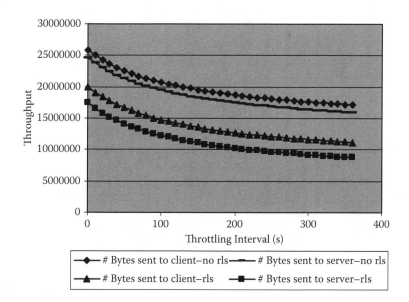

Figure 16.8   Impact of throttling interval on throughput. (Source: Nokia-Siemens Networks.)

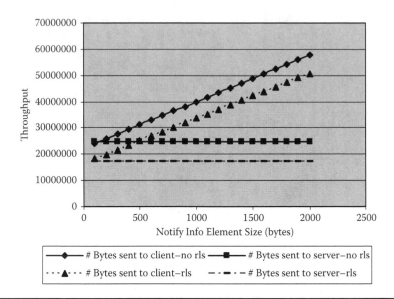

Figure 16.9 Impact of NotifyInfoElement size on throughput. (Source: Nokia-Siemens Networks. Used with permission.)

subscriber (watcher) is querying a large group of contacts, if a contact's presence status has not changed, there is no value in sending that information. Only contacts whose presence status has changed will be sent, and it is up to the client to present the total view of the service to the subscriber. Thus, standards have defined two optional functions to limit the size of a presence message:

1. Partial publication: The presentity (or presence source) publishes only the parts of its information that have changed since it sent its last PUBLISH message to the Presence Server.
2. Partial notification: The Presence Server sends to the watcher only the changes in the presence status information since its last notification. The changes can be reflected on a presentity basis (as in the address book example given earlier) or within a presentity (such as a change in the mood indicator).

One other method is available to reduce the size of the message over the air interface. This involves compressing the message using signaling compression algorithms such as SigComp or DEFLATE. These compression techniques typically require a more powerful device processor. Until recently, these processors were only available in the most expensive, high-end mobile devices. As the price–performance curve drops, it is expected that all mobile devices will have the processing power needed to support compression. It should be noted that SigComp is only used between the

P-CSCF and the mobile device. Within the network, the message is required to be decompressed for the message headers to be read and interpreted.

Endnotes

1. As discussed in Chapter 3, although the preliminary foundation work for Presence was done in IETF and 3GPP, by agreement OMA has been given the responsibility for the standardization of enablers for the IMS, including Presence.
2. Even though standards has defined the interface between the PNA and the HLR (i.e., P_h reference point), most carriers will not implement this interface, because they would not want to burden such a mission-critical platform with these types of queries. Instead, a more creative solution would be for the PNA to interface to a probe on the SS7 link connected to the HLR. Here, the messages going to and from the HLR can be monitored through the probe to determine if a subscriber has registered or not.
3. Of course to some watchers, "Polite Blocking" could be viewed as "Impolite." Use of this privacy setting indicates that a watcher is not allowed to see the presentity's status information; also that watcher is not even allowed to know that he or she is actually blocked.
4. This last "×2" multiply is because for each SIP message sent, there is a corresponding acknowledgement (i.e., 200 OK message) sent, which doubles the traffic load.

Further Reading

Open Mobile Alliance, "Presence Requirements," version 1.0, OMA-RD-Presence _SIMPLE-V1_0.

Open Mobile Alliance, "Presence SIMPLE Architecture Document," version 1.0.1, OMA-AD-Presence_SIMPLE-V1_0_1.

Open Mobile Alliance, "Stage 2—Presence Using SIMPLE," version 1.1, OMA-PAG-SIMPLE-AD-V1_1.

Open Mobile Alliance, "Presence SIMPLE Specification," version 1.0.1, OMA-TA-Presence_SIMPLE-V1_0_1.

3GPP TS 22.141, "Presence Service; Stage 1."

3GPP TS 23.141, "Presence Service; Architecture and Functional Description."

3GPP TS 24.141, "Presence Service Using the IP Multimedia (IM) Core Network (CN) Subsystem; Stage 3."

RFC 1951, "DEFLATE," Deutsch, P., May, 1996.

RFC 3261, "SIP: Session Initiation Protocol," Rosenberg, J., Schulzrinne, H., Camarillo, G., Johnston, A., Peterson, J., Sparks, R., Handley, M., and E. Schooler, June 2002.

RFC 3265, "Session Initiation Protocol (SIP)-Specific Event Notification," Roach, A. B., June 2002.

RFC 3320, "Signaling Compression (SigComp)," Price, R., Bormann, C., Christoffersson, J., Hannu, H., Liu, Z., and J. Rosenberg, January 2003.

RFC 3863, "Presence Information Data Format (PIDF)," Sugano, H., Fujimoto, S., Klyne, G., Bateman, A., Carr, W., and J. Peterson, August 2004.

RFC 4119, "A Presence-Based GEOPRIV Location Object Format," Peterson, J., December 2005.

RFC 4480, "RPID: Rich Presence Extensions to the Presence Information Data Format (PIDF)," Schulzrinne, H., Gurbani, V., Kyzivat, P., and J. Rosenberg, July 2006.

RFC 4589, "Location Types Registry," Schulzrinne, H. and H. Tschofenig, July 2006.

RFC 5112, "Presence-Specific Static Dictionary for Signaling Compression," Garcia-Martin, M., January 2008.

Chapter 17

XML Document Management Server

Many of today's applications are intelligently designed to have flexible service logic that would take customer- and service-specific data into consideration when processing a service request. Consider the popular Instant Messaging (IM) service as an example; here, we will investigate what is the customer- and service-specific data that could be used when processing an IM service request.

Alice has recently received a new challenging video game and is trying to advance to the next game level. Her friends Bob, Cheryl, Dave, Emma, and Frank are also into this new video game, and each of them is also trying to advance to their next game level. Alice would like to chat with them from time to time to exchange experiences and challenges she faces with this new video game. Because all her friends have their own IM service, she would like to use IM to chat with them. Alice invokes her mobile IM service from her mobile IM service provider to start a chat with Bob, Cheryl, Dave, Emma, and Frank. It is a wonderful conversation, and Alice learns some new tricks from them to advance to the next game level. A few days later, Alice struggles to advance to yet another game level and she thinks she might get some hints from her friends again. So, she asks her mobile IM service to start another chat with Bob, Cheryl, Dave, Emma, and Frank. This periodic chat among Alice and her friends continues for some period of time until Alice beats the game. Just imagine what a bad user experience Alice would have if every time she started an IM chat with her friends she needed to type in all her friends' IM addresses. A better user experience for Alice would be if she could just select her chat friends from her device address book and start the IM chat from there. The best user experience for Alice would be to create an IM chat group for this topic with her friends Bob, Cheryl, Dave, Emma, and Frank as group members and use this group to start a chat every time she needs to. In this case, she just needs to remember the predefined IM chat group name (she could use a name that is meaningful to her and also easy to remember) and use it, instead of typing in all the members IM addresses to start a group chat.

So, Alice asks her mobile IM service provider to create a chat group, *A-Game*, for her, which is illustrated in Figure 17.1.

The IM service processes the request from Alice (step 1), creates an *A-Game* chat group that is stored (step 2) for later use, and acknowledges to Alice the creation of the A-Game group (step 3).

Later, Alice would like to start a chat with the A-Game group members and asks the IM service to set it up, as illustrated in Figure 17.2.

When the IM service processes Alice's request to start a group chat with *A-Game* members (step 1), it first retrieves the *A-Game* group members from storage (step 2) and then uses it to set up connections with all members for a group chat (step 3 and 4).

This example demonstrates the need to store some service-specific data in order to process service requests. The service-specific data mentioned in the previous example is created by a user (Alice), and is stored in the network by the application

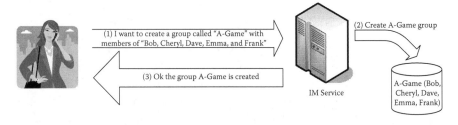

Figure 17.1   IM chat group creation.

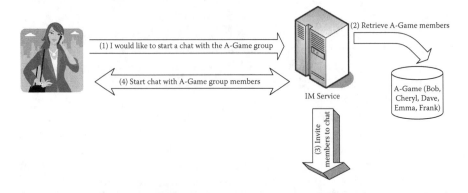

Figure 17.2   Start a predefined IM chat.

(IM service) to help process service requests by an application (IM service). There could be other customer- and service-specific data for an IM service that will make the service even more attractive to a user, with even better user experience, such as Alice's profile data, her access policy, and group messaging management, etc.

A different type of service-specific data is created by an application while processing a service request, which could be managed by the user and used for later service request processing. Take the example of the Presence[1] service. The service might maintain a list of the watchers who are authorized to get the updates of a particular Presentity's presence information. The Presence service updates this list every time it processes an authorization request, and the Presentity can obtain access to this list to know who is making a request for his or her presence information.

We will use Open Mobile Alliance (OMA) relevant specifications to explore the customer- and service-specific data in details in the following sections.

17.1 OMA XDM Enabler

OMA first identified the need to manage and access the customer- and service-specific data during the development of the Push-to-Talk over Cellular (PoC) Service Enabler[2] in 2003. PoC Service Enabler supports predefined group PoC communications, and a PoC user needs to be able to manage (such as create, edit, retrieve, and remove, etc.) the predefined PoC groups. To support this need, OMA started the "Group Management" enabler development in January 2004, using text documents to manage the PoC predefined group list information. It was developed as an OMA-supporting enabler for other service enablers to use as well. The name of the enabler was changed to "eXtensible Markup Language (XML)[3] Document Management" (XDM) Enabler in late 2004 to better represent its generic and reusable nature. Figure 17.3 shows the XDM Enabler development history.

As illustrated in Figure 17.3, the XDM Enabler development was spawned from the PoC Enabler development in 2003 and was started in 2004. The XDM Enabler version 1.0 development was completed in April 2006, and it supports only PoC Enabler version 1.0. The XDM Enabler version 2.0 development was started in 2005, after the completion of the XDM Enabler version 1.0 technical specifications development; its technical specifications development was completed in 2007 and is currently under the validation process.[4] XDM Enabler version 2.0 supports both PoC Enabler version 2.0 and SIMPLE IM Enabler version 1.0. The XDM Enabler version 2.1 development was started in 2007, after the completion of XDM Enabler version 2.0 technical specifications development; it is targeted to be completed in 2009. XDM Enabler version 2.1 will support PoC Enabler version 2.1 and Converged-IP Messaging (CPM) Enabler version 1.0. The XDM Enabler version 1.1 development was started in 2007 and completed in 2008; it fixes a number of technical issues in XDM Enabler version 1.0 specifications.

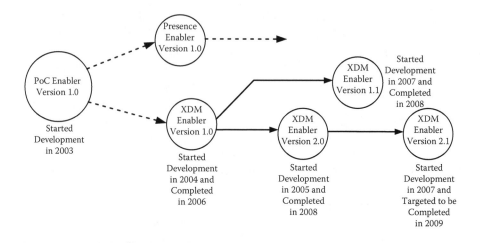

Figure 17.3   OMA XDM architecture development history.

17.2 Why XML Documents?

Data can be stored in many different formats: ASCII text, UTF-8 text, binary, bitmap just to name a few. To select the right data format to store the customer- and application-specific data, OMA was looking for the following characteristics:

■ It needs to be created and managed by a customer, so it needs to be text based and readable by humans.
■ It needs to be accessed by an application, and easy to understand and process.
■ It needs to be efficiently managed in a resource-limited environment.

Text is the most easily understood and widely used data format. A markup is a set of annotations to text that describe how it is to be constructed, laid out, or formatted and is widely used in the printing industry. It has been used since ancient time and, in recent years, it has also been used in computer typesetting and word-processing systems. It is used directly over the Internet to exchange documents in an expected form. XML is a general-purpose specification developed by the World Wide Web Consortium[5] (W3C) for creating customer markup languages. XML is extensible because it allows users to define their own elements in a tree structured like a document. It is used to communicate richly structured documents between applications and is human readable.

A software module called an *XML processor* is used to read XML documents and provide access to their content and structure. An application could use the XML processor to process XML documents. The XML processor is easy to create and is widely available today. Considering all the benefits of an XML document,

OMA has decided to use it to store and manage the customer- and service-specific data in the form of XML documents.

So, when Alice wants to create the *A-Game* chat group, her device would actually create an XML document called *A-Game* and send it to the IM service in the network, and the IM service would store it as an XML document named *A-Game* in the network storage. Later, when Alice requests the IM service to start a chat with the *A-Game* group, the IM service would retrieve the *A-Game* XML document from the storage and extract the member information from the document to invite them to join the chat with Alice. The sample XML document of *A-Game* would look like this:

```
<?xml version="1.0" encoding="UTF-8"?>
<resource-lists xmlns="urn:ietf:params:xml:ns:resource-lists"
  <list name="A-Game">
  <display-name>A Video Game</display-name>
    <entry uri="sip:bob@att.net">
      <display-name>Bob</display-name>
    </entry>

    <entry uri="sip:cheryl@att.net">
      <display-name>Cheryl</display-name>
    </entry>

    <entry uri="sip:dave@att.net">
      <display-name>Dave</display-name>
    </entry>

    <entry uri="sip:emma@att.net">
      <display-name>Emma</display-name>
    </entry>

    <entry uri="sip:frank@att.net">
      <display-name>Frank</display-name>
    </entry>

  </list>
</resource-lists>
```

A human could easily read this XML document and know that the group being created is called *A-Game group*. The display name for this group is "A Video Game." The members of the *A-Game* group are Bob, Cheryl, Dave, Emma, and Frank, as shown by their display names.

The customer- and service-specific data is stored as XML documents. The repositories in the network where the XML documents are stored are called *XML Document Management Servers* (XDMSs). The XDMS is a logical representation of the repository of XML documents, and is located in the network for easy access.

17.3 OMA XDMS Architecture

OMA XDM Enabler is based on the popular client-server model. The XDM client (XDMC) resides in an end-user device or is embedded in an application depending on the usage, whereas the XDM servers are located in the network. The XDM Enabler architecture has evolved from version 1.0, which supports only PoC Enabler, to version 2.0, which supports both PoC and SIMPLE IM Enablers. We will focus on XDM Enabler version 2.0 architecture in our discussions.

Figure 17.4 shows the major components in the OMA XDM Enabler architecture. The XDMC communicates with the XDM Server either directly or indirectly to search and manipulate XML documents. The aggregation proxy server provides device-resident XDMC easy access to multiple XDM servers in the network, and the search proxy server provides the XDMC the ability to look up specific information in XDM servers.

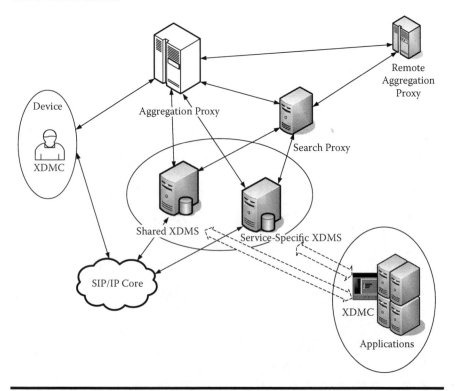

Figure 17.4   OMA XDM architecture.

17.3.1 XDM Client

There are two kinds of XDMC: a device-resident XDMC and an application-embedded XDMC. The device-resident XDMC provides an interface for a user to manipulate XML documents in the XDMS; in our example, Alice uses it to create an *A-Game* group. The application-embedded XDMC provides an interface for an application to access the needed XML documents in the XDMS while processing a service request; in our example, the IM service uses it to retrieve Alice's A-Game group to establish an IM chat group for her.

The device-resident XDMC also provides a different interface to the search proxy server in the network to support a user searching relevant information in the network.[6] It supports a subset of XQUERY[7] functions over the Hypertext Transfer Protocol (HTTP)[8] communications link with the search proxy server, through the aggregation proxy server, for search operations. The supported XQUERY functions are "request" and "response."

Based on the user requests or the application needs, the XDMC constructs an XML Configuration Access Protocol (XCAP)[9] URI to access and manage the XML documents in the XDMS. XDMC needs to be authenticated before it is allowed to access the XDMS. The device-resident XDMC is authenticated by the aggregation proxy server, whereas the application-embedded XDMC is authenticated directly by the XDMS. The application-embedded XDMC is configured with the XDMS address (i.e., SIP URI) for direct access to the XDMS.

17.3.2 Aggregation Proxy Server

To prevent potential malicious attacks, mutual authentication is performed before a XDM client-server communication is established; so, before an XDMC could access or manage an XML document with XDMS, a mutual authentication is performed. To start communications with the XDMS, the XDMC needs to know where that XDMS is located (i.e., the SIP URI of the XDMS). In some operations (such as in "search"), the XDMC might access multiple XML documents stored in different XDMSs and needs to know the addresses of these XDMSs to perform mutual authentication and to establish communication. It would be resource intensive and time consuming because the XDMC needs to know the location of all XDM servers and might need different credentials to authenticate with each one of them.

The aggregation proxy server in the network eases this task by providing a single point of entry into the XDMS for the services it provides. This simplifies the authentication process[10] for the XDMC to access all relevant XDM servers and consolidates the results from different XDM servers into a single reply message back to the XDMC. The aggregation proxy server also supports XDMC accesses to remote XDM Servers offered by different service providers with a secured communication link if there is a service level agreement in place. The aggregation proxy server also supports data compression and decompression to make efficient use of radio bandwidth. In general, the aggregation proxy server provides the XDMC efficient access to the XDM servers.

17.3.3 Search Proxy Server

It would be a good user experience if a user could know what information is being stored in XDM servers; it could also provide efficient use of valuable radio bandwidth to manage only the XML documents that are needed. The OMA XDMS architecture provides this function through the search proxy server in the network. Search proxy server was added in OMA XDM Enabler version 2.0.

The search proxy server provides a single point of entry to support XDMC search requests. A user may want to search the XML documents that he or she created or the XML documents others created which he or she is allowed to search for information. In our example, Alice might want to search the IM community public profiles that her IM service provider supports to see if there are other game lovers who could provide her hints to play her video game. To simplify the authentication process and the presentation of search results, the XDMC communicates with the search proxy server through the aggregation proxy server. The search proxy server will analyze the search request coming from the XDMC and forward it to the appropriate XDM servers to process. The search proxy server will gather the search results from XDM servers and provide a single response back to the XDMC. With the interoperability and service level agreement in place, the search proxy server could provide XDMC the ability to search remote XDM servers or other non-XDM directories[11] to get better results and enrich the user experience.

A subset of XQUERY functions over the HTTP[12] communications link is used for search operations. The supported XQUERY functions are *request* and *response*.

17.3.4 Service-Specific and Shared XDMS

XDMS is the repositories of XML documents in the network. Based on the scope of the service-specific data and its usage, OMA has divided the XML documents into two categories: *service-specific* and *shared*.

The service-specific XML documents are data that are only meaningful to and accessed by that specific service enabler. When Alice records an IM conversation with the A-Game group so that she could later refresh her memory, the IM service will store the conversation content in a local storage and its related metadata in the IM-specific XDMS. The metadata, in the form of URI, is used to retrieve and manipulate the stored conversation history contents. Figure 17.5 illustrates the IM-specific XDMS. The service-specific XDM server supports the search operations.

On the other hand, the *A-Game* group list that Alice created for IM service could also be used by other services. For example, if all of Alice's friends also have the PoC service, then the *A-Game* group could potentially be used by the PoC service to set up a PoC call among the *A-Game* members. The OMA shared XDMS is the repository for XML documents that could be shared by multiple services. Based on the XML document's characteristics, the shared XDMS is currently divided into four logical servers: Policy XDMS, List XDMS, Group XDMS, and Profile XDMS. Figure 17.6 gives the high-level illustration of the shared XDMS.

The shared *Policy XDMS* is the repository for the user access policy XML documents. It defines access rules for a user/application in the form of condition-action pairs according to RFC 4745.[13] The condition part of an access rule is used to

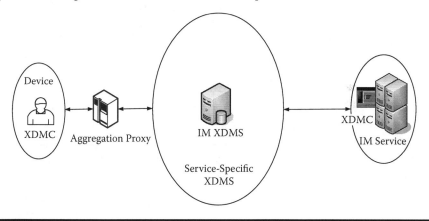

Figure 17.5   Service-specific XDMS.

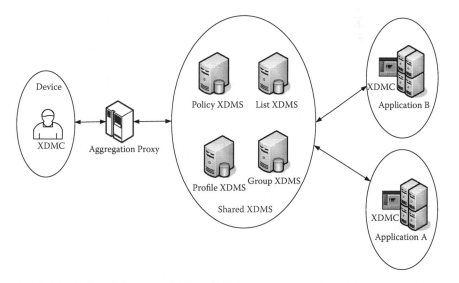

Figure 17.6   OMA shared XDMS.

match the request, and the action part of the access rule is used to determine which operations should be applied. In a way, an access rule could be read as "if (*condition*) then (*action*)." An example of an access rule is for Alice to define "block all communications from calls with no caller-ID."

The shared *List XDMS* is the XML document repository for the following two types of shared list XML documents:

- List of URI XML documents that could be used for service request processing, such as a list of individual URIs that could be used to block communications (e.g., blacklist).
- List of group URI XML documents that used to describe the characteristic of a defined "group." Example of information captured in this XML document includes description of the group, media that will be used, and supported services.

The shared *Group XDMS* is the repository for the group XML documents. Such documents are used for managed rules in group conversations. In our earlier example, when Alice creates *A-Game* group, the group rules information will be stored here. It includes information related to a "group" such as the name of the group, the subject of the group, who are invited, how many people are in a group, access rules, media allowed, supported services, etc.

The shared *Profile XDMS* is the repository of the XML document of user profiles. The data in a user profile is divided into two logical parts: unlocked-user-profile and locked-user-profile. The unlocked-user-profile is controlled and managed by a user, and may contain personal information such as name, mailing address, birth date, gender, hobbies, etc. The locked-user-profile is controlled and managed by the service provider, which may contain the more accurate "official" user's birth date.[14] For an age-restricted service (such as an IM chat room for adults 18 years and older), the service could check the user's locked-user-profile[15] to determine if the user is allowed to use that service.

The shared Profile XDMS supports the *search* operation. An authorized user could search the shared Profile XDMS to look up a person matching his or her search criteria. In our example, Alice could search the shared Profile XDMS to see if there are other gamers matching her search criteria whom she could exchange experiences with. Only the unlocked-user-profile part of data in the shared Profile XDMS is searchable.

These shared XDM servers could work together to provide a better service. Let us consider the example of Bob trying to get access to Alice's profile, as illustrated in Figure 17.7.

After the application accepts Bob's request to access Alice's profile (step 1), it will first check with Alice's access policy through Alice's shared Policy XDMS

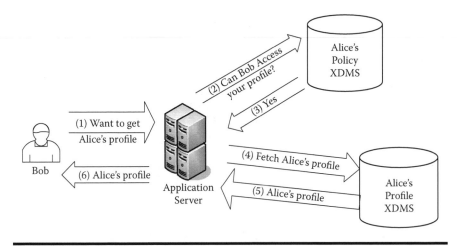

Figure 17.7   Bob accesses Alice's profile.

(steps 2 and 3), and only fetch her profile XML document from Alice's shared Profile XDMS (steps 4 and 5) if authorized by Alice. The application then sends Alice's profile to Bob (step 6) to complete the service request.

17.4 XML Configuration Access Protocol

The device-resident XDMC supports two interfaces to the aggregation proxy server, one to support the management of XML documents and the other one to support *search* operations. To support the management of XML documents, OMA has adopted the IETF-developed *XML Configuration Access Protocol (XCAP)* for communications between XDMC and XDMS; hence, an XML document stored in XDMS is constructed as an XCAP Application Usage.[16]

In general, the Application Usage defines the following:

- Application Unique ID (AUID), which is uniquely identified within the application usage namespace. The namespace could be an IANA-registered IETF namespace or a vendor-proprietary namespace.
- XML schema.
- MIME type.
- Validation constraints.
- Data semantics.
- Authorization policies.

OMA XDMS defines XCAP User Identifier (XUI) to access the XML document in the network. The XUI is a string of path elements in an HTTP URI that is associated with each user served by the XDMS. It is in the generic form of

```
[XCAP Root URI]/[AUID]/users/[XUI]/...
```

The XCAP Root URI points to the aggregation proxy server of the XDMC's home domain. As an example, the following XCAP URI is used to access Alice's profile:

```
GET /org.openmobilealliance.user-profile/users/sip:alice
  @att.net/user-profile HTTP 1.1
Host: att.net
```

To make the protocol simple, XCAP only supports the following meaningful HTTP operations:

- *GET Operation*: XDMC uses GET operation to retrieve an XML document, element, or attribute from XDMS.
- *PUT Operation*: XDMC uses PUT operation to create or modify an XML document, element, or attribute with XDMS.
- *DELETE Operation*: XDMC uses DELETE operation to remove an XML document, element, or attribute from XDMS.

Figure 17.8 illustrates the abstract diagram when Alice accesses the members in the *A-Game* chat group.

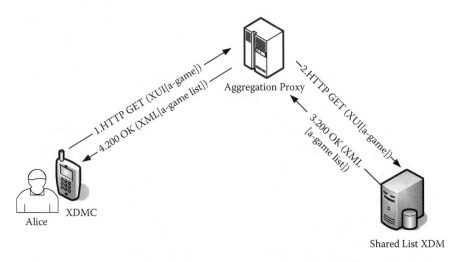

Figure 17.8 Alice retrieves *A-Game* group list.

After mutual authentication with the aggregation proxy server, the XDMC on Alice's device constructs an XUI (using the AUID of the shared List XDMS, Alice's personal information, and Alice's service request) to retrieve the *A-Game* group members' information. It then initiates the HTTP GET operation with the shared List XDMS, via the aggregation proxy server (step 1). The aggregation proxy server passes the HTTP GET operation to the target XDMS, the shared LIST XDMS to process (step 2). The shared LIST XDMS processes the XUI and uses the path information to retrieve the XML document that contains *A-Game* group and returns it to the aggregation proxy server (step 3). The aggregation proxy server forwards the resulting XML document to XDMC to process (step 4). The XDMC processes the XML document and presents a user-friendly format to Alice.

17.5 Subscribe to XML Document Changes

Subscribing to the changes in an XML document is an efficient way to keep up with the current content of an XML document. This operation is especially valuable in a resource-limited environment such as in a mobile network. Alice could subscribe to the changes to her IM conversation history XML document, and she would be notified if there is a new conversation history being recorded. She could then use the notified information, a link to the conversation history file, to retrieve that specific conversation history.

17.5.1 Subscribe and Notify

SIP SUBSCRIBE and NOTIFY methods[17] [RFC 3265] are used to allow the XDMC to keep up-to-date information about an XML document stored in the XDMS. Figure 17.9 illustrates how Alice would get the notification of changes to her IM conversation histories from the IM XDMS.

Alice subscribes for the changes to her IM histories XML document in the IM service-specified XDMS through the SIP/IP Core (steps 1–4). Alice then starts an IM chat with her *A-Game* group and requests the IM service to record this conversation (step 5). Once the conversation ends, the IM service stores this recorded conversation metadata into Alice's histories XML document in the IM XDMS (step 6). The IM XDMS sends a notification, with the appropriate information, to Alice through the SIP/IP Core (steps 7 and 8). Alice could store the information received from the notification and use it to retrieve the recorded conversation history. The SUBSCRIBE and NOTIFY are SIP methods, and this exchange transverses the SIP/IP core and does not go through the aggregation proxy server.

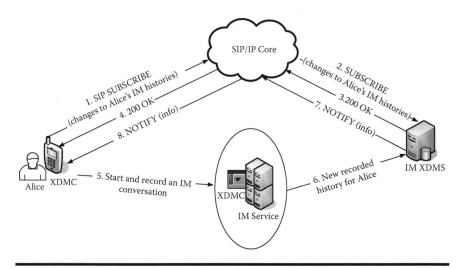

Figure 17.9   Sample subscribe-notify flow.

17.6 Future of OMA XDM Enabler

The OMA vision of XDM Enabler is a supporting enabler. It is evolved based on the needs to support other enablers' development. OMA is currently developing XDM Enabler version 2.1 to support both PoC Enabler version 2.1 and CPM Enabler version 1.0. The XDM Enabler version 2.1 is targeted to be approved as an OMA Enabler in later 2009. OMA-Converged Address Book (CAB) Enabler[18] development has identified the XDM server as a potential repository for a network-based address book; thus, further XDM Enabler development in the future will be needed. There are currently several products on the market that support XDM Enablers today. As more OMA Enabler-based services are deployed, we will see more products supporting OMA XDM Enabler in the marketplace.

Endnotes

1. See Chapter 16 for more information about Presence service.
2. Refer to Chapter 18 for PoC Service Enabler.
3. Refer to Chapter 13 for more information on XML.
4. Refer to Chapter 3 for more on the OMA Enabler development process.
5. The W3C is an international consortium that develops Internet Web standards and guidelines. It has published more than 110 Web standards that are called *W3C recommendations*.
6. The searched information could be in the XDMS or external to the XDMS such as the white page and yellow page databases.
7. XML Query is a query language for XML documents developed by the W3C.

8. HTTP is a communications protocol, developed jointly by the IETF and the W3C, used to transfer information on the Internet and the World Wide Web.
9. Refer to Chapter 13 for details about XCAP.
10. Once the aggregation proxy server authenticates the XDMC, it will share the XDMC identity with XDM servers and the Search Proxy.
11. If supported by a service provider, a user could search a white page database, yellow page database, or an enterprise name directory to find the information he or she needs.
12. HTTP is a communications protocol, developed jointly by the IETF and the W3C, used to transfer information on the Internet and the World Wide Web.
13. RFC 4745 defines a rule consisting of three parts: conditions, actions, and transformations. The shared Policy XDMS only uses the first two parts in setting rules.
14. Normally, the service provider will enter locked-user-profile birth date information during the provisioning process with validated data such as from the user driver's license.
15. The birth date information in the unlocked-user-profile is not to be trusted since it is managed by the user.
16. An XCAP Application Usage, in the context of OMA XDM Enabler, is the detailed information on the interaction of an application with XDMS.
17. SIP method is a well-defined "request" message in a SIP request-response transaction.
18. OMA CAB Enabler is targeted to be approved as an OMA enabler in mid-2009.

Further Reading

OMA, "XDM Document Management," version 2.0, http://www.openmobilealliance.org.

RFC 4745, "Common Policy: A Document Format for Expressing Privacy Preferences," Schulzrinne, H., Tschofenig, H., Morris, J., Cuellar, J., Polk, J., and Rosenberg, J., February 2007.

IETF RFC 2616, "Hypertext Transfer Protocol," Fielding, R., Gettys, J., Mogul, J., Frystyk, H., Masinter, L., Leach, P., and Berners-Lee, T., June 1999.

XQuery: An XML Query Language, http://www.w3.org/TR/xquery/.

IETF 3265, "Session Initiation Protocol—Specific Event Notification," A. B. Roach, June 2002.

Chapter 18

Push-to-Talk over Cellular

Most of the telecommunications systems we use today are full-duplex communications systems; a typical example is a landline (fixed-line) telephone system. A full-duplex communications system allows two parties in communication to communicate in both directions at the same time.

Figure 18.1 shows User A and User B engaging in a communication with two communications channels. One communications channel is employed by User A for transmission and User B for reception. The other communications channel is employed by User B for transmission and User A for reception. In this arrangement, both User A and User B can talk (i.e., transmit) and listen (i.e., receive) at the same time.

To duplicate the same user experience as in a landline telephone system, the conventional mobile communications system is designed in a similar way, providing a full-duplex communications system to allow its users to have simultaneous conversations in both directions. Figure 18.2 illustrates this scenario. Similar to a landline full-duplex system, the mobile network allocates two radio frequencies (i.e., two communications channels) for a conversation between User A and User B. One radio frequency is allocated to allow User A to transmit and User B to receive and another radio frequency is allocated for User B to transmit and User A to receive.

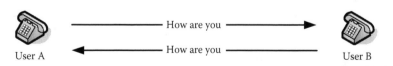

Figure 18.1   Full-duplex communications.

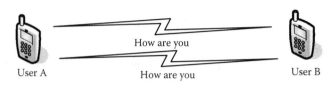

Figure 18.2 Wireless full-duplex communications.

The communications can also be carried in a half-duplex communications system. In a half-duplex communications system, there is only one communication channel allocated for both parties to communicate. The communication channel can either be used for sending or receiving speech (i.e., media), but not for sending and receiving at the same time. Thus, when one party is speaking (i.e., transmitting), the other party can only be listening (receiving). A transmission collision happens if both parties transmit at the same time. To avoid transmission collision, a method of regulating who has the right to use the communication channel to transmit needs to be in place.

Figure 18.3 illustrates User A and User B in a half-duplex communications system. When User A is talking (i.e., transmitting), User B would only be listening (receiving). User B could only "talk" when User A finishes "talking" and releases the use of the communication channel. A typical half-duplex communications system is the walkie-talkie device that many of us experienced as a toy when we were young. There is a button on the walkie-talkie device that a user has to push to start transmission (i.e., talk); thus, half-duplex communications has been known as *Push-to-Talk (PTT) communications.*

One distinct characteristic of PTT communications is that you do not have to "dial" the party you would like to talk to; you simply push the talk button and speak into your PTT device. The recipient party on the other end will hear your voice instantly if his or her walkie-talkie device is turned on. Another important feature of PTT communications is its broadcasting nature, in that multiple walkie-talkie devices will receive the same transmission at the same time if they are all tuned to the same radio frequency.

The PTT concept has been successfully developed into commercial communications services over the years and widely used in the dispatching industries.[1] Nextel Communications™ (now a part of the Sprint Nextel Corporation™) offers the most successful PTT communications services in the marketplace. All the commercial PTT communications services offered today are based on proprietary solutions. Interoperability between two PTT communications service providers based on two different vendors' solutions is very challenging, if not impossible.

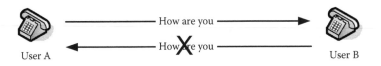

Figure 18.3   Half-duplex communication system.

18.1 OMA PoC Enabler

Seeing the success of commercial PTT communications services and anticipating the huge market opportunities, Open Mobile Alliance (OMA) started a work item on Push-to-Talk over Cellular (PoC) in 2003.[2] The goal of this work item was to develop a service enabler to offer PTT-like (i.e., half-duplex) communications services with rich features and interoperability. The first release of OMA PoC Enabler, version 1.0, was completed in 2006. The technical specifications development for a later release, OMA PoC Enabler version 2.0, has been completed and is under validation. OMA is now working on OMA PoC Enabler version 2.1 technical specifications development.

18.1.1 OMA PoC Versions

18.1.1.1 OMA PoC Enabler Version 1.0

OMA PoC Enabler version 1.0 focuses on development of the basic PTT-like voice communications framework. The main features supported in OMA PoC Enabler version 1.0 are:

- Half-duplex voice communications for one-to-one or in a group as a PoC session[3]
- Floor control procedures in a PoC session
- Automatic or manual answer mode
- Personal alert
- Group list management[4]
- Presence support

To accommodate the nature of the half-duplex communications characteristics (i.e., only one party in a PoC session can talk at any one time), a floor control procedure was developed to regulate who in a PoC session has the rights to transmit (i.e., speak). We will explore the floor control mechanism in greater detail in Section 18.1.7.

OMA PoC Enabler supports both *automatic* and *manual* answer modes to accommodate different user expectations. For instant communications, a PoC user could set the PoC answer mode to automatic, and the incoming voice communications will be played right away at the recipient's device without the need for action on the part of the recipient. There will be occasions when a PoC user is in an environment in which privacy is needed but not provided. In this situation, he could set the PoC answer mode to manual. In the manual mode setting, an incoming PoC session will need to be answered by the recipient before a PoC session can start.

Typically, when you visit a colleague in his office and he is not there, you would leave him a note to tell him that you have visited and may even ask him to call you when he sees the note. The OMA PoC Enabler offers the *personal alert* feature that gives a PoC user this capability. The *personal alert* feature allows a PoC user to leave an indicator with the unanswered party that informs them that he or she should call back.

OMA PoC Enabler supports group conversations both in an ad hoc group and a predefined group. A PoC user could start an ad hoc PoC session by including all the invitees in the invitation. A PoC user could set up a PoC group by using the group list management feature and later start a PoC session with members in this predefined group. The group list management feature is provided by OMA XDMS Enabler. A specific PoC group is designated as a *PoC Dispatch Group*, with unique group rules to support dispatch services.

Optionally, OMA PoC Enabler supports presence information of PoC users. This feature is provided with the support of the OMA Presence Enabler.

18.1.1.2 OMA PoC Enabler Version 2.0

Whereas OMA PoC Enabler version 1.0 provides a framework to support half-duplex voice communications services, OMA PoC Enabler version 2.0 focuses on enriching the PoC user experience. Some of the main features supported by OMA PoC version 2.0 are

- Enriched voice call experience with multimedia contents
- Discrete messaging capability (IM-like)
- Interworking with existing PTT services
- Machine-to-user PoC session
- Third-party-initiated PoC session
- PoC Box (to store media bursts)

A plain voice conversation could be greatly enriched by including multimedia content such as video and image. OMA PoC Enabler version 2.0 supports a separate communication channel to exchange multimedia contents in a PoC session. The popular instant messaging service is also supported in a PoC session. Thus, OMA PoC Enabler version 2.0 could be used to offer Push-to-X services such as Push-to-See, Push-to-Video, and Push-to-Share, etc.

To maximize the OMA PoC service reach, it is important to interwork with the existing PTT services. This makes it possible for a PoC session to include participants who are either OMA PoC service users or existing PTT service users. This interworking capability also provides a smooth transition for the existing PTT service providers to migrate into open-standards-based OMA PoC service.

Instead of supporting human-only PoC sessions for voice communications in OMA PoC Enabler version 1.0, the OMA PoC Enabler version 2.0 supports extending a PoC session to a machine (such as a PC) or an application (such as a recording service). A PoC Box provides media storage for a PoC user and works in a similar way to how a voice mail box works today. With a PoC Box, a PoC user would never miss a PoC session. We give a detailed description of PoC Box in Section 18.1.8.

18.1.2 OMA PoC Enabler Architecture

OMA PoC Service Enabler is built with the popular client-server model, shown in Figure 18.4. The client provides user experience to a PoC user while the server supports PoC features.

The high-level OMA PoC architecture in Figure 18.4 only shows the major components that are the building blocks for OMA PoC Enabler framework. The core components are the PoC Client in the device and the PoC Server in the network; together they provide all capabilities needed to establish PoC sessions for communications. The XDMS[5] in the network is used to store OMA PoC Enabler-related data in a PoC group session such as predefined group list information. The Presence Server[6] in the network provides presence-related information on a PoC user and the contacts in his or her predefined group lists.

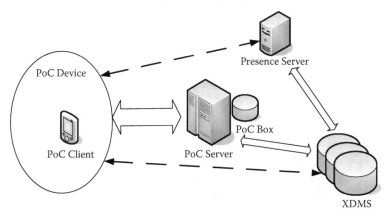

Figure 18.4   High-level view of OMA PoC architecture.

18.1.3 OMA PoC User Experience

The PoC end user experience is provided through the PoC device user interface, interworking mainly with the PoC Client. Figure 18.5 provides an illustrative view of the OMA PoC user experience. The user experience could be further enriched by coordinating with other OMA-supporting Enablers such as XDMS and Presence.

The use of OMA XDMS Enabler is crucial to support PoC group communications where the predefined group lists information is stored. The use of OMA Presence Enabler is optional, but if used would further enhance the end-user experience.

The OMA Device Management (DM) Enabler is used to provision the PoC Client settings.

18.1.4 OMA PoC Client

The PoC Client in a PoC device communicates with PoC Server in the network using specific protocols for various PoC features. The PoC Client, through the specific PoC device user interface, reacts on a PoC user's action and provides subscribed PoC services. The PoC Client would authenticate a PoC user before allowing the user to use PoC services such as initiating a PoC session. To provide different user experiences and manage different expectations, the PoC Client supports the use of the Quality of Experience[7] (QoE) profile when initiating a PoC session.

During a PoC session, the PoC Client sends and receives media bursts, depending on the PoC user's action and the media floor control procedures, and then presents the results to the PoC user (such as transmitting the PoC user's speech or

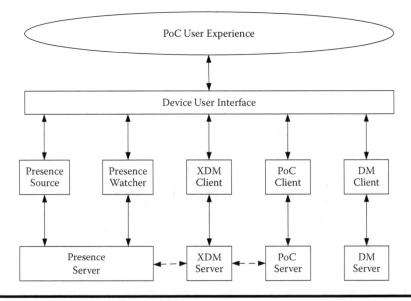

Figure 18.5   OMA PoC user experience.

presenting the received speech from others). With OMA PoC Enabler version 2.0, the PoC Client also supports nonspeech discrete media communications.

Different protocols are used between PoC Client and PoC Server to meet their needs. SIP is used mainly to perform authentication, authorization, and PoC session management. RTP[8] is used for real-time media transport, and RTCP[9] is used for Talk and Media Burst control. MSRP[10] is used for Discrete Media transportation.

18.1.5 PoC Server

The PoC Server is the core network element that provides application-level network functions for the PoC service. To better define the role a PoC Server plays in different PoC session scenarios, the core PoC Server functions are divided into two logical entities: the *PoC Controlling Function* and the *PoC Participating Function*. The PoC Participating Function maintains a direct communications link to a PoC Client that services the requests to/from a PoC user. The PoC Controlling Function does not have a direct communications link to a PoC Client; instead, it maintains a direct communications link to each of the PoC Participating Functions that a PoC Client connects to. Figure 18.6 depicts the relationship between the PoC Controlling Function, the PoC Participating Function, and a PoC Client.

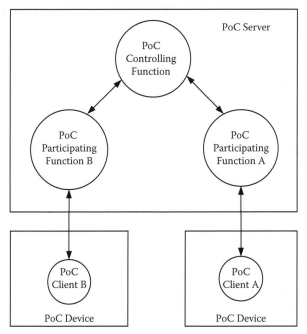

Figure 18.6 OMA PoC Server functions.

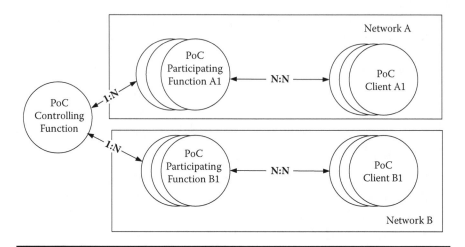

Figure 18.7   Relationship between PoC Controlling Function, PoC Participating Function, and PoC client.

Figure 18.6 shows a PoC session established between PoC Client A and PoC Client B. The session connection is always going through the PoC Participating Function which serves the connecting PoC Client in a one-to-one relationship. The two PoC Participating Functions are also connected through the PoC Controlling Function in a one-to-one relationship. The location of the PoC Controlling Function varies, depending on the deployment options. Normally, the PoC Controlling Function resides in the domain of the service provider of the PoC Client that initiates the PoC session. In every PoC session, there is always a PoC Server and only one performing the PoC Controlling Function, and a number of PoC Participating Functions corresponding to the number of participating PoC Clients, as illustrated in Figure 18.7.

In Figure 18.7, there are three participants (A1–A3) in network A and three participants (B1–B3) in network B in a PoC session. There are three PoC Participating Functions in Network A; each one manages the PoC session for one of the participating PoC Clients, such as PoC Client A1 connecting to PoC Participating Function A1 to join in this PoC session. Similarly, there are three PoC Participating Functions in Network B managing three different PoC Clients for this PoC session. The PoC Controlling Function could be located in Network A or Network B, depending on which PoC client initiates the PoC session; or it can be located in a third network, Network X, as a dedicated network that offers this PoC session.

In a PoC Server, the following functions are performed by the PoC Participating Function:

■ Session management
■ Policy enforcement

- Charging information collection
- Floor control
- Interworking with non-OMA PTT networks
- Content adaptation and traffic optimization
- Support QoE

In a PoC Server, the following functions are performed by a PoC Controlling Function:

- Centralized session management and media distribution
- Floor control
- PoC group sessions and dispatch PoC sessions
- Discrete media
- PoC Box handling
- Traffic optimization
- Support QoE for each session

A PoC user is assigned a SIP Uniform Resource Identifier (URI)[11] as a PoC address to be used to communicate with other PoC users. A traditional phone number (i.e., Directory Number) could also be used as a PoC address, but it needs to be converted into a SIP URI by the PoC Server before it can be used for setting up a PoC session. A PoC group identity that is used in a PoC group conversation also takes the form of a SIP URI.

In addition to half-duplex voice speech (i.e., Talk Bursts[12]), OMA PoC Enabler version 2.0 also supports the following media types in a PoC session:

- Audio (e.g., music)
- Video
- Discrete media (e.g., still image, text, and file)

The Talk Burst Control mechanism developed for OMA PoC Enabler version 1.0 has been evolved into Media Burst control mechanism in OMA PoC Enabler version 2.0 with backward compatibility (refer to Section 18.1.7 for more information).

18.1.6 Discrete Media Support

OMA PoC Enabler version 2.0 supports discrete media transfer in a way that is similar to how the Instant Messaging service provides it today. The discrete media types supported for transfer in a PoC session are *text, audio, video, image,* and *text file.* The IETF MSRP[13] protocol is used to transfer the supported discrete media. In addition to setting up a PoC session just for discrete media transfer, an existing PoC session could also be modified to transfer discrete media. A discrete media transfer report could be requested by the initiating PoC Client to obtain the transfer status and ensure the media transfer is completed.

18.1.7 Media Burst Control

Due to the half-duplex nature of the PoC service, a floor control mechanism is needed to regulate a PoC session; that is, to authorize who has the rights to transmit in a PoC session and for how long. A *talk burst* is the flow of media from a PoC Client while it has the permission to send media. The only media supported in OMA PoC version 1.0 Enabler is half-duplex voice communications (i.e., talk bursts); hence, the Talk Burst Control mechanism was developed. The main purpose of the Talk Burst Control mechanism is to define a set of protocols to regulate a PoC session on the following:

- How to request to transmit talk bursts
- How to handle multiple requests (i.e., queuing and preemption)
- How to inform PoC session participants on incoming talk bursts
- How to release the rights to transmit talk bursts

To handle multiple requests, a queuing mechanism is developed with several priority levels. A *high priority* request is granted before a *normal priority* request. A PoC client with *listen-only* priority cannot request to transmit Talk Bursts. A request with *preemptive priority* will be granted with rights to transmit talk bursts immediately; if someone else is transmitting talk bursts at that moment, his rights will be revoked unless that person is also transmitting with "*pre-emptive priority.*"

With multimedia supported in OMA PoC Enabler version 2.0, the Talk Burst Control mechanism has been evolved into Media[14] Burst Control mechanism. The Media Burst Control mechanism supports all the capabilities the Talk Burst Control mechanism supports, with additions for floor control. A *media floor control* is a mechanism to control separate media streams in a PoC session.

A state machine called *Media-floor Control Entity* is developed to ensure that only one participant in a PoC session can access the media resource at the same time. Both the PoC Client and PoC Server support Media-floor Control Entity and the characteristics of media types. Media-floor Control Entity is negotiated by the PoC client and the PoC server during a PoC session establishment.

In OMA PoC Enabler version 2.0, a PoC session might consist of one or more media streams. It is also possible to have one or more media streams for the same media type. For those media streams that use the Media-floor Control mechanism, each media stream or multiple media streams should be controlled by separate Media-floor Control Entities as illustrated in Figure 18.8.

Figure 18.8 shows the binding of Media-floor Control Entity with "half-duplex voice," "video and audio," and "audio" media types. Text is not bound to any of the

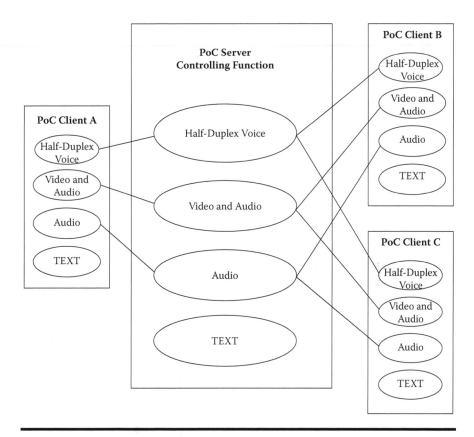

Figure 18.8   PoC Media-floor Control mechanism.

Media-floor Control Entities, because no floor control is needed for discrete media transfer. The PoC Participating Functions are not shown in Figure 18.8, because they are not significant in this scenario.

18.1.8 OMA PoC Box

The PoC Box provides similar functions as a voice mailbox today, to record a PoC session for a PoC user. The PoC Box could reside in the network as network-based storage or in a PoC device as device-resident storage. The use of PoC Box is optional and is based on a PoC user's settings. The OMA PoC Enabler supports three different PoC Box usage settings: *never, conditional,* or *unconditional.* The PoC user is able to change these settings at any time. The *never* setting means a PoC Box will never be used during a PoC session.

If a PoC user wants to use PoC Box for an incoming PoC session, the PoC Participating Function that connects to the PoC Client would route the session to the PoC Box. The network-based PoC Box should always be used to alleviate the

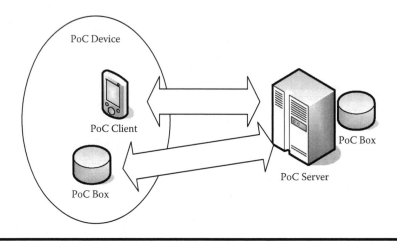

Figure 18.9 OMA PoC Box.

limited PoC device storage. The device-based PoC Box will be used only if the PoC user has defined such usage; PoC Box can only be used in a PoC group session with more than two participants.

In Figure 18.9, both device-resident PoC Box and network-based PoC Box are shown. If the device-based PoC Box is used (instead of a PoC Client) in a PoC session, it behaves like a PoC Client receiving all the incoming media bursts and storing it. The network-based PoC Box provides similar functionality as a device-based PoC Box in a PoC session.

18.1.9 Quality of Experience (QoE)

In a real-world scenario, PoC users will have different needs for network resources, and a PoC service provider could apply the quality of experience (QoE) profile at subscription time to meet the user's expectations. The QoE profile is used to map expected QoE of PoC users for the applications they intend to use. The QoE profile, when applied, enables end-to-end quality-of-service management capability to meet the PoC user's expectation. OMA PoC Enabler supports four levels of QoE profiles for different user expectations:

- **Basic:** The lowest QoE profile level, intended for a PoC session with no special QoE expected; normally means "the best-effort" service.
- **Premium:** For users with high expectations of demanding QoE, such as in an interactive communications service.
- **Professional:** For PoC sessions with professional applications that demand certain level of QoE, such as in a multimedia streaming service.
- **Government:** For users who require priority access to PoC services according to the five levels defined in the Wireless Priority Service [RFC 4412]. The

use of this profile is subject to local and national regulations. The user of this profile enjoys precedence over all other QoE profiles.

The QoE profile maps the quality of service (QoS) to be provided for each media type in a PoC session. A PoC user could indicate to the PoC Controlling Function the desired QoE profile at the initiation of a PoC group session. The PoC Controlling Function will assign a unique QoE for this specific PoC session after the authorization is checked by the serving PoC Participating Function. If a PoC session is initiated without a QoE profile, the "basic" QoE profile will be assigned. The QoE profile cannot be changed during a PoC session.

Preemption is supported when "government" QoE profile is used. When it happens, if a PoC client does not support multiple PoC sessions, the current PoC session will be released and the new PoC session will be started.

18.1.10 OMA PoC Dispatcher

The dispatching service is one of the most prevalent PTT types of communications services offered today. OMA PoC Service Enabler supports this type of service through *PoC Dispatcher* functionality as a special prearranged PoC group session called *Dispatch PoC Group*.

There are two roles for all participants in a Dispatch PoC Group: *PoC Dispatcher* and *PoC Fleet Member* (Figure 18.10). One or more members in a Dispatch PoC Group could be assigned the role of PoC Dispatcher, but only one

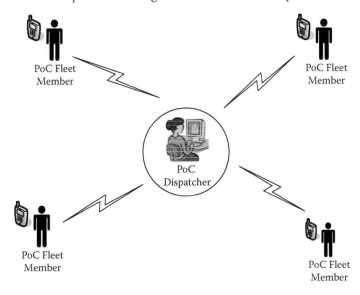

Figure 18.10 OMA PoC Dispatch Service.

is active at any given time in a Dispatch PoC session. The members who are not assigned the role of PoC Dispatcher are in the role of PoC Fleet Members. If a member who is assigned the PoC Dispatcher role is not active in this role, he or she will act as a PoC Fleet Member.

The PoC Dispatcher is the focal point in a Dispatch PoC session, having the right to expel any PoC Fleet Members if necessary; all media exchange during a Dispatch PoC session is between a PoC Dispatcher and PoC Fleet Member. There are no direct communications channels among PoC Fleet Members in a Dispatch PoC session.

The Dispatch PoC session is normally initiated by a group member who has been designated the PoC Dispatcher role. The Dispatch PoC session could be started with all members or a subset of members of that specific Dispatch PoC Group. A PoC Fleet Member could invite himself or herself to an existing Dispatch PoC session or initiate a dialog with one of the PoC Dispatchers if no Dispatch PoC session exists.

A PoC Dispatcher could transfer his or her role to another PoC user who also has been assigned a role as PoC Dispatcher. If the PoC Dispatcher role has changed, all PoC members in that Dispatch PoC Group will be notified.

18.1.11 Interwork with Other Legacy PTT Networks

To expand the coverage of OMA PoC service, especially in the early stage of OMA PoC service deployment and to enhance the OMA PoC user experience, the interworking with existing PTT services is needed. A PoC Interworking Function is developed to interwork with the existing PTT service. Both the signaling messages and media flows will be converted by the PoC Interworking Function as illustrated in Figure 18.11.

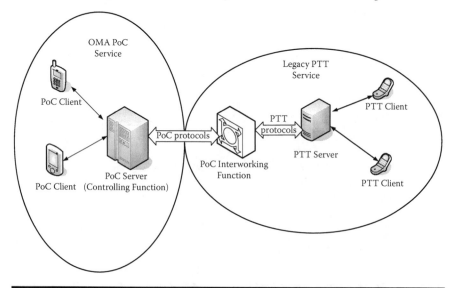

Figure 18.11   OMA PoC and legacy PTT interworking.

Figure 18.11 shows that the PoC Interworking Function is used to connect the OMA PoC Server and legacy PTT Server. The PoC Interworking Function could be located either in the OMA PoC service domain or in the existing PTT service domain.

18.2 OMA Supporting Enablers

Traditionally, a service enabler development is always self-contained; that is, all the features required for that service enabler are developed. It is called *vertical development model*. An example of the vertical development model is illustrated in Figure 18.12.

In Figure 18.12, three service enablers, X, Y, and Z, are being developed. Service Enabler X requires features A, B, and C. Service Enabler Y requires features A and D. Service Enabler Z requires features B, E, and F. In a vertical development model, each service enabler is self-contained and independent of other enablers, so feature A will be developed when Service Enabler X is developed and it will be developed again when Service Enabler Y is developed. Similarly, feature B will be developed for Service Enablers X and Z separately. The main advantages in a vertical development model are that (1) all features required for a service enabler are developed tightly together, which makes the interfaces among those features simple and (2) easier control of development schedule. The disadvantages of a vertical development model are (1) the duplicated development efforts and (2) the potential inconsistent solutions for same features developed for different service enablers. The use of the vertical development model is also known to create a silo effect.

A different development model is called the *horizontal development model*. In the horizontal development model, common features among service enablers are

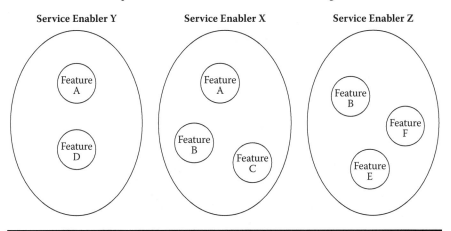

Figure 18.12   A typical vertical development model.

identified during the development process, and these common features are developed as common enablers to be used by needed service enablers. An example of the horizontal development model is illustrated in Figure 18.13.

The same three service enablers (X, Y, and Z) developments are shown in Figure 18.13. Features A and B have been identified for use in more than one service enabler, and thus are separated out as independent development efforts and provide standardized interfaces for service enablers that need to use their features. In this case, Feature A is used by Service Enablers X and Y through a standardized interface, and feature B is used by Service Enablers X and Z through a standardized interface. There are no redundant efforts to develop common features, and the development of common features across different service enablers is consistent. When there is an enhancement or change to a common feature, it only needs to be developed once and applied to all service enablers that use it. The challenges of a horizontal development model are to (1) identify what are the common features, (2) determine how to standardize their interfaces, and (3) coordinate development schedules among service enablers and common features.

Seeing the benefits of the horizontal development model, OMA has decided to adopt it in its service enabler's development. When the PoC Enabler was under development, the "group list management" and "presence" features were identified as potential common features that could be used by other service enablers.[15] It was decided at that time to separate these two features from PoC Service Enabler development. The XDMS Enabler was created to support the "group list management"

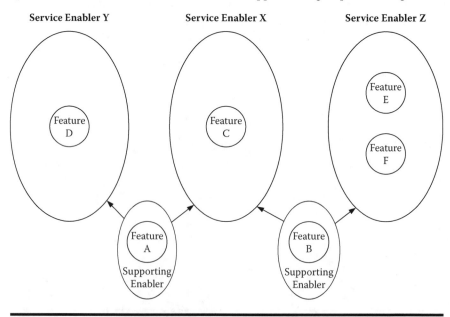

Figure 18.13   A typical horizontal development model.

feature, and the Presence Enabler was created to support the "presence" feature. OMA has designated this kind of enabler as *supporting enabler*, and the XDMS and Presence are supporting enablers of OMA PoC Service Enabler.

Other than being identified as potential common features during a service enabler development such as XDMS and Presence, some common functionality has been developed to play the role of a supporting enabler for other service enablers. OMA DM Enabler is a good example. It provides functionality for a network operator to provision a device to be used for certain services. The supporting enablers for PoC Service Enabler are shown in Figure 18.5.

18.3 Future of OMA PoC Enabler

When OMA started the PoC Enabler development in 2003, it attracted a lot of attention and was one of the biggest development efforts in OMA; at the height of its development it attracted more than 100 delegates from most of the major companies in the IT and Telecom industries around the world to its meetings. It is still developing a new version of PoC Enabler[16] today to support even more features. The thousands of technical input contributions to the OMA PoC Enabler's development proved to have a major impact on other OMA Service Enablers' development. The PoC architecture model, especially the division of Participating Function and Controlling Function, has been adopted in the OMA SIMPLE IM and OMA CPM Service Enablers' development.

Endnotes

1. A trucking industry, a cab-dispatching service, and police communications are all typical dispatching services.
2. The term PoC is meant to designate a standardized version of the walkie-talkie service PTT (i.e., the OMA version). PTT has a connotation with a proprietary version of the walkie-talkie service. Because of its previous widespread use, PTT is used by many in the industry as a generic term for the service and is used (albeit incorrectly) interchangeably.
3. A PoC session is set up for communication between two parties or in a group environment.
4. Refer to Chapter 17 for more information.
5. Refer to Chapters 13 and 17 for more information.
6. Refer to Chapter 16 for more information.
7. See Section 18.1.9 in this chapter for details of QoE.
8. Real-time Transport Protocol is an IETF-developed protocol to support real-time services.
9. Real-time Transport Control Protocol is an IETF-developed protocol that provides out-of-band control for RTP flows.

10. Refer to Chapter 14 for more information.
11. A SIP URI is defined in IETF RFC 2396 and used by SIP users to communicate with each other. See Chapter 10 for more details.
12. A talk burst is the flow of media from a PoC client while it has the permission to send media.
13. See Chapter 14 for more details on MSRP.
14. In a sense, media covers any user data transmitted over a PoC session, including Talk Bursts.
15. The OMA SIMPLE IM Service Enabler was under development at that time and would require these two similar features to support it. Later on, OMA CPM Service Enabler would also use these two features.
16. The latest version is OMA PoC version 2.1, which was started in August 2007 and is scheduled to be completed in early 2011.

Further Reading

1. OMA, "Push to Talk over Cellular (PoC)," version 1.0, http://www.openmobilealliance.org.
2. OMA, "Push to Talk over Cellular (PoC)," version 2.0, http://www.openmobilealliance.org.
3. IETF RFC 4412, "Communications Resource Priority for the Session Initiation Protocol (SIP)," H. Schulzrinne and J. Polk, February 2006, http://www.ietf.org/rfc/rfc4412.txt.
4. OMA, "Device Management," version 1.2, http://www.openmobilealliance.org.

Chapter 19

IP Messaging

Many readers will be familiar with the Instant Messaging (IM) services that are provided by popular Internet portals. IM services have historically been PC-based, although they have begun to penetrate the wireless domain.

For purposes of this book, IP Messaging refers primarily to IM-like services that run over Internet Protocol networks. The term *IP Messaging* signifies a departure from wireless-specific protocols such as Short Message Service (SMS). SMS is the workhorse for messaging traffic in wireless networks today, but it is not IP-based. Although the term "IP Messaging" comes from the wireless domain, it is a convergence play. By and large, fixed-line networks do not understand SMS. On the other hand, fixed-line broadband access is predominantly IP-based.

19.1 IP Messaging, Presence, and XDM

The IP Messaging service we describe in this chapter is essentially that defined by the Open Mobile Alliance in OMA SIMPLE IM. It leverages the Presence service defined by OMA which is described at length in Chapter 16. OMA SIMPLE IM also leverages OMA's XDM enabler (e.g., to manage contact lists). Chapter 17 is dedicated to the XDM Enabler.

19.2 Taxonomy of Messaging Types from OMA SIMPLE IM

OMA SIMPLE IM defines three messaging types: pager mode, large message mode, and session mode. We further classify pager mode and large message mode

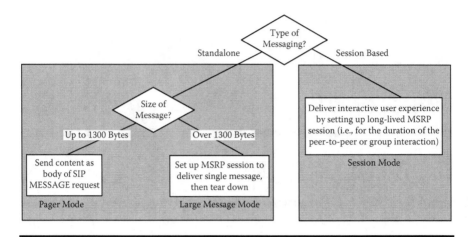

Figure 19.1   OMA taxonomy of messaging types.

as standalone messaging, because neither provides a true session-based user experience. Figure 19.1 illustrates the standalone versus session-based dichotomy.

OMA coined the term *pager mode* to evoke a functionality comparison with two-way pagers. (IETF and Third Generation Partnership Project [3GPP™] call it "page mode," but in our opinion that term is somewhat confusing.)

Large message mode is invoked whenever the content of a standalone message will not fit in a SIP MESSAGE. In Chapter 11, Section 11.9.6, we noted that the maximum size for a SIP MESSAGE—or, for that matter, any SIP request—is 1300 bytes. This is the reason that large message mode "starts" at 1300 bytes. In large message mode, a unidirectional Message Session Relay Protocol (MSRP) session is set up for the purpose of transferring a single message. Once the transfer is complete, the session is torn down. Session mode is used to present a true session-based experience to the end user.

Large message mode is unique to OMA—-neither IETF nor 3GPP defines it. Why does OMA distinguish between session mode and large message mode? This is because the SIP MESSAGE size limitation seems artificial to end users who do not understand the reason for the limitation. In order to enable a seamless stand-alone messaging user experience (represented by the shaded box on the left in Figure 19.1), OMA has defined large message mode and grouped it with pager mode. Said differently, end users may very well be aware of the distinction between stand-alone and session-based messaging, but will not be aware that the former is comprised of two modes.

19.3 Pager Mode

Multiparty messaging is supported in each of the three OMA-defined messaging modes. Figure 19.2 depicts a multiparty use case involving multiparty messaging.

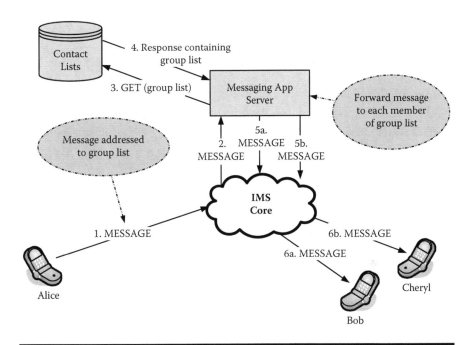

Figure 19.2   Pager mode messaging.

We assume that when the signaling flow of Figure 19.2 starts, there is already a group list belonging to Alice whose members are Bob and Cheryl. This list is similar to the *A-Game* group list of Section 17.2 (the contents of the XML document are displayed in that section). The only difference is that the *A-Game* list contains three more entries: Dave, Emma, and Frank. We have scaled down to two entries here to keep the diagram simple.

In Figure 19.2, Alice sends a SIP MESSAGE to the aforementioned group list. In the IMS core, the S-CSCF forwards the MESSAGE to the Messaging Application Server based on a *feature tag* that appears in the SIP MESSAGE header. For OMA SIMPLE IM, the feature tag is the character string "+g.oma.sip-im"; we will not concern ourselves with exactly where the feature tag is situated in the SIP MESSAGE.

The Messaging Application Server is responsible for fetching the group list identified in Alice's MESSAGE. The cylinder labeled "Contact Lists" in the figure is an XDM server as described in Section 17.3. The arrows numbered 3 and 4 in Figure 19.2 depict an XCAP operation running over Hypertext Transfer Protocol (HTTP). Details are suppressed here in the interest of simplicity.

Let us suppose, as in the *A-Game* example of Section 17.3, that Alice shares an interest in multiplayer gaming with Bob and Cheryl. The use of pager-mode messaging, as depicted here, would be suitable for an interchange where Alice wants to find out if Bob and Cheryl are available to play the following night.

19.4 Session Mode

Figure 19.3 depicts a messaging session involving the same protagonists as the previous section. Alice addresses a message to the same group list, but this time it is an `INVITE`. As before, the S-CSCF directs the `INVITE` to the Messaging Application Server based on the presence of the appropriate feature tag. The Messaging Application Server is then responsible for retrieving the group list and sending an `INVITE` to each list member. The IMS core and Contact Lists XDM server are omitted here in order to keep the diagram simple, but they function in very much the same fashion as in the example of Figure 19.2.

Once the messaging session is set up, MSRP messages can flow between the participants. The figure shows an MSRP message—sent by Alice—being broken into chunks as described in Section 14.2. The Messaging Application Server is responsible for forwarding messages to both recipients. In this example, Bob and/or Cheryl can also send messages to the other participants within the context of the same session. We simply ran out of room to portray multidirectional exchanges in a single diagram.

The use case depicted in this section would make sense when Alice, Bob, and Cheryl are engaged in a multiparty gaming session. In this case, the players' devices are all running at least two applications (the gaming application itself and

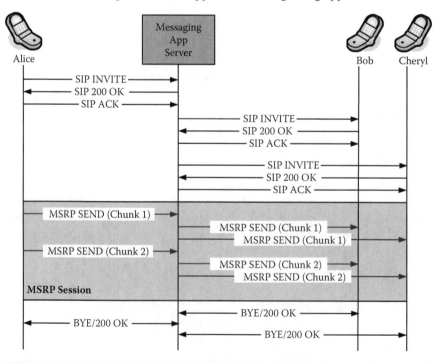

Figure 19.3 Multiparty messaging session.

the messaging application just described). Multiple applications may very well be sharing one connection to the network. Any message containing a large attachment should not be allowed to "clog" the connection at the expense of gaming traffic—especially if game performance is highly sensitive to delay. In such an environment, MSRP's chunking capability is an asset.

19.5 Large Message Mode

In Section 19.2, Alice sends a message to a group list whose members are Bob and Cheryl. The implementation given there (see Figure 19.2) uses SIP MESSAGE. One would expect it to be rare for a text message to exceed the capacity of SIP MESSAGE. As noted in Section 19.2, a SIP MESSAGE can be up to 1300 bytes in length.

What if Alice wants to attach a picture to her message, but does not want to enter into a full-fledged messaging session at the moment? In use cases such as this, the 1300-byte limit will routinely be exceeded. This is the motivation for large message mode. Rather than draw a new diagram, we approach large message mode in terms of similarities and differences with session mode.

In session mode described in Section 19.4, any of the session participants could originate an MSRP message. This is a typical case for session-based messaging, although the specifications are flexible enough to allow for cases where some participants can receive messages but do not have permission to send messages.

Figure 19.3 could just as easily depict a use case for large message mode. The difference is that the MSRP session is required to be unidirectional—only the participant who originates the INVITE is allowed to send. Once the session setup is completed, a single message is sent and the session is immediately torn down. This does not preclude MSRP chunking, however. In large message mode, it is important to note that the end users are unaware that an MSRP session setup takes place.

19.6 Deferred Messaging and History Function

What happens when a stand-alone message is sent to a user whose device is not attached to the network? Perhaps the device is turned off or is out of coverage. In this case, the Messaging Application Server queues the message. The message is delivered whenever the intended recipients' devices are available to receive messages. Upon successful delivery, messages are deleted from the aforementioned queue.

Since OMA SIMPLE IM is a presence-enabled service, why would such cases arise? Even if the sender is maintaining presence information on the recipients, that presence information could be out of date. But there is another reason that is arguably more important: interoperability with non-IMS messaging. In particular,

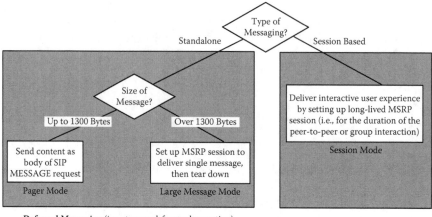

Figure 19.4 Applicability of deferred messaging and conversation history function.

many "legacy" wireless devices have clients for SMS and Multimedia Messaging Service (MMS). There is no guarantee that a presence-enabled IP Messaging user will have access to presence information about users of legacy devices.

As indicated in Figure 19.4, deferred messaging is applicable to the two standalone messaging modes—namely, pager mode and large message mode. Deferred messaging does not provide long-term storage in the network; as noted earlier, messages are deleted from the Messaging Application Server's queue upon successful delivery.

If the intended recipient of a deferred message does not attach to the network for an extended period of time, how long should that message be retained? This is typically an operator configurable setting. An operator must weigh the cost of deferred message storage versus customer satisfaction in receiving the message. The Messaging Application Server must maintain a queue for deferred messages so that each such message can be quickly and efficiently delivered upon receipt of notification that the intended recipient has registered on the network. Because of the implied fast access requirement, a deferred message queue is more expensive to operate (on a per-megabyte basis) than a data warehouse. A data warehouse is just what it sounds like—a mass archival storage facility. In current practice, the maximum storage duration ranges from a couple of days to a couple of weeks. Once a message exceeds the maximum duration, it is cleared from the message queue.

OMA SIMPLE IM also defines a *conversation history function*. Compared to deferred messaging, there are two major differences:

- The conversation history function applies to all three messaging modes.
- The conversation history function is related to long-term storage of messages.

The last point needs some clarification. The conversation history function stores and manages metadata in an XDM server, along with pointers to the messages in question. Thus, the messages themselves can be housed in another repository such as a data warehouse. In this case *metadata* is information—e.g., when the messaging interchange occurred and who was involved—that facilitates later searching of archives.

In the wireless domain, message history has traditionally been relegated to the end user's device. That is, the only persistent storage for past SMS and MMS messages resides on the wireless device. The conversation history function utilizes network-based storage instead. Why this departure from tradition? One of the main reasons is that it supports multidevice access.

Figure 19.4 summarizes the OMA taxonomy of messaging types that was first presented in Section 19.2. It also shows the applicability of deferred messaging and the conversation history function in that context.

19.7 File Transfer

OMA SIMPLE IM also supports use cases requiring file transfer capabilities. OMA SIMPLE IM merely leverages the functionality covered in Chapter 14, Section 14.4, which describes the use of MSRP to support file transfer applications.

19.8 Converged IP Messaging

OMA SIMPLE IM provides a foundation for IP Messaging. It defines a presence-enabled service that supports group scenarios for standalone as well as session-based messaging types. As we saw in Section 19.6, SIMPLE IM anticipates multidevice access requirements by incorporating a conversation history function. However, SIMPLE IM does have significant limitations in this and other areas.

Like SIMPLE IM, OMA's Converged-IP Messaging (CPM) is a SIP-based messaging framework. Moreover, CPM has a broader charter than that of SIMPLE IM. CPM's broadened scope for IP Messaging includes the following:

- Extended multidevice functionality. For instance, the CPM requirements call for the ability to transfer a CPM session from one device to another device. As an example, a user may begin a CPM session on a broadband PC in the home. If the user needs to leave the house but wishes to continue the CPM session, he or she can transfer the session to a mobile device.
- The ability to leverage capabilities of value added services within the context of a CPM session. For example, suppose a group of friends or colleagues are

engaged in a group messaging session and are deciding where they should meet for dinner. One of the participants invokes a restaurant finder application, which sends a list of suggested restaurants, along with information about each restaurant, to all of the session participants. Once a restaurant has been selected, a mapping application is invoked to send directions to each session participant.

- Interworking with non-CPM messaging services, including SMS, MMS, and SIMPLE IM.
- Seamless interworking with a network-based address book.
- Support for multiple CPM addresses (for the same CPM user). For example, a CPM user might have a work CPM address that is shared with colleagues and a personal address that is shared with friends and relatives.
- Presence settings can differ from one CPM address to another. One effect of the aforementioned multi-address capability is the potential for friends and relatives to see one collection of presence settings (e.g., available to communicate during evenings and weekends) and colleagues to see a different collection of presence settings (unavailable during evenings and weekends).
- Last, CPM's scope includes sessions with continuous media, discrete media, or a mix of the two. The underlying philosophy is that conference calls, Push-to-Talk over Cellular sessions (see Chapter 18), and discrete messaging sessions should all be built on a common framework. Returning to the example in which a group of people selects a restaurant, it should not matter which sort of session the participants choose to engage in; basic functionality should be as consistent as possible.

Further Reading

OMA, "SIMPLE IM," version 1.0, http://www.openmobilealliance.org.

OMA, "Converged IP Messaging Requirements," candidate version 1.0, http://www.openmobilealliance.org.

RFC 4975, "The Message Session Relay Protocol (MSRP)," Campbell, B., Mahy, R., and Jennings, C., September 2007.

APPLICATIONS AND USE CASES

Chapter 20

IMS Ecosystem

The rapidly changing competitive landscape for the telecommunications industry is forcing both incumbent operators as well as new entrants to look for new ways to quickly differentiate themselves from other competitors. This is evident when looking at traditional voice pricing, which was once the hallmark of the industry and is now widely offered as a commodity priced service. Thus, operators must now seek new and innovative services to replace traditional voice revenues, as well as to provide growth revenue opportunities.

Not to be overshadowed, new waves of competitors are entering the once-protected domain of the incumbent telecom operator. From cable companies to Internet companies to new technology entrants (e.g., WiMAX, satellite, etc.), all are looking to gain market share away from the established operators. All parties involved recognize that in order to win or to maintain market share, new and innovative services will need to be deployed in a rapid, time-to-market fashion in order to maintain competitiveness.

The traditional long lead time for deployment of a new service will need to become a footnote from the past as operators move toward a rapid service deployment model and an efficient life cycle management process in order to remain competitive. Niche services with a specialized target market will offer one way for operators to open new revenue streams. Here, time-to-market will be critical in order to stay competitive and to capture a greater market share. This new rapid service deployment paradigm will look to reuse different service components already deployed to meet time-to-market requirements as well as to reduce overall deployment costs in order to cost justify the new service.

The incorporation of the IP paradigm and the control mechanisms that IMS provides will enable new classes of services not currently possible with today's circuit-switch telephony networks. Combined with higher access bandwidths, IMS is expected to open up new spectrums of services to benefit end users as well as to provide new streams of revenue opportunities for network operators and end-users' service providers.

In this final part of the book, we want to focus on how a network operator/service provider's marketing organization might approach a decision to deploy a particular service. Typically, this process begins with an understanding of a demand or a problem scenario that is not being addressed. They will then try to clearly articulate the need through identifying real-world scenarios and specifying them in use case scenarios. Next, they should be getting together with their technology or engineering organizations to brainstorm how the needs can be best met in a timely and economical manner. By understanding the new classes of services that IMS enables, service designers can apply those solutions to their marketing's organization needs. Finally, after the brainstorming or ideation phase completes its iteration, then the design of the service is begun.

The next three chapters present some hypothetical use case scenarios that a marketing organization might develop and bring to their engineering organization for service design. Based upon what we have learned in the first three parts of the book, solutions are then constructed to solve the different use case scenarios. It should be noted that the flexibility of IMS can allow for different but equally valid implementations of a service (different from those presented in the following chapters). Many factors must be considered when designing an actual service implementation.

Before we get to our use case scenarios, we want to look at two topics in this chapter. First, we will look at some examples of different classes of services that can be blended together so that IMS will enable the network operators and service providers to monetize their investment in IMS. Second, we want to briefly introduce an industry effort to enable the development of new applications in a responsive, time-to-market manner. Here we want to introduce the concept of a Service Delivery Framework.

20.1 IMS Classes of Services

There are many ways to group telephony applications together, depending upon your perspective since no formal taxonomy exists. In this section, we will discuss several well-known or even de facto classes of services that IMS enables. Knowing the capabilities that an IMS network can enable will allow service designers to determine how to resolve a particular communication problem.

Before we proceed, it might be helpful to provide a distinction between the terms *service* and *application*. Although sometimes they are used as interchangeable descriptors, there are specific differences. A *service* is what is offered/sold to the end user or customer. It can consist of a single or multiple applications using certain capabilities[1] of the network with a charging plan applied. An application is an atomic unit providing one particular function (or feature). Confusion in terminology occurs when a service is made up of only a single application and thus that application is also a service (and, depending upon context, called a service).

It might be easiest to explain the difference with an example. Let's use as an example service the Video Share service. It consists of the application's two-way voice call (which is a common service by itself) plus one-way video streaming. It can be charged as normal air-time (for a mobile call) with either a per video invocation charge or a data throughput charge. It requires certain capabilities of the network such as the ability to exchange SIP messages to set up the session and a broadband wireless access (e.g., UMTS™/HSPA, EV-DO Rev. A) network. Thus, one of the key benefits of IMS is its ability to blend multiple applications into new end-user services in a much easier (translate less costly) and quicker manner.

For our purposes, let us classify applications into one of three categories:

1. Real time[2]
2. Interactive
3. Background

20.1.1 Real Time

Real-time applications have latency requirements measured in hundredths of milliseconds and are usually defined as not noticeable by the end user. The most common real-time applications are associated with making or receiving a voice call. In a pure IMS network, this would be a Voice-over-IP (VoIP) call. Existing supplementary services associated with a basic voice call such as call waiting, call forwarding, etc., are being defined for deployment in an application server designated in standards as an MMTel (Multimedia Telephony) Application Server.

Another example of a real-time service is a live streaming feed. As shown in Figure 20.1, a video-streaming AS can be invoked for the purpose of content

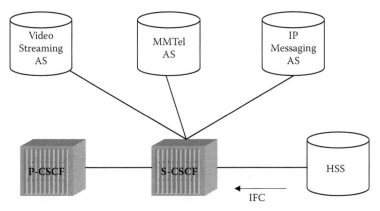

Figure 20.1   Real-time architecture applications examples.

adaptation or to record the live video stream into the customer's personal storage locker. Now by blending a voice call with a live streaming feed, one can create the multimedia (voice media plus streaming media) service known as Video Share. In our example, both the voice path and video streaming path are combined into a single bearer session to create the multimedia service. Supplementary services can be invoked during the video share session by invoking features on the MMTel AS. As other real-time application servers are added, these can be combined using either the iFC or service broker function to do application blending.

Before we leave our discussion on real-time application, it is appropriate to discuss one difference between GSM-based and CDMA-based networks when it comes to voice and IMS. When GSM carriers adopted the UMTS evolutionary path for their 3G networks, they adopted a technology that inherently could support simultaneously a voice path plus a data path. This allows for blending the two bearers immediately from initial deployment so services such as Video Share can be deployed immediately. In contrast, CDMA carriers selected EV-DO (EVolution-Data Only) for their 3G technology.[3] While EV-DO supports broadband access rates similar to UMTS, a simultaneous voice connection cannot be supported with a data application. Thus, in order for a CDMA carrier to offer a voice call connection with a data application, they must migrate their network and devices to support VoIP first. While VoIP will be the long-term method for offering voice services over a cellular carrier's network, in the near term the spectral efficiencies of either (UMTS or EV-DO) technology to support VoIP are inferior to the spectral efficiencies of a 3G circuit switch voice call. Thus, GSM carriers do not have the financial incentive in the near term to move to VoIP in the near future, especially since they have an alternative method to offer simultaneous voice–data applications. On the other hand, CDMA carriers will have to migrate to VoIP sooner than may be financially justifiable if they wish to deploy simultaneous voice–data applications. As the technology evolves, it is expected that VoIP will pass the spectral efficiencies of a 3G circuit-switched call and thus will be adapted by the GSM operator community.

20.1.2 Interactive

An interactive service can be thought of similar to a browser-type service in which data information flows between a server and a client. Here, an end user can query a server for a particular piece of information or action. This can vary from the well-known Internet browser session, to modifying one's profile preference list, to updating the presence status for members of an address book group list.

An interactive service can be blended with a real-time service or with other interactive services. This is made possible as IMS can support multiple sessions in context for the end user. So, for example, while Bob is on a call with Alice and trying to get from his location to hers, he can query a location server to find Alice's location and then browse to a mapping server to obtain turn-by-turn directions (Figure 20.2).

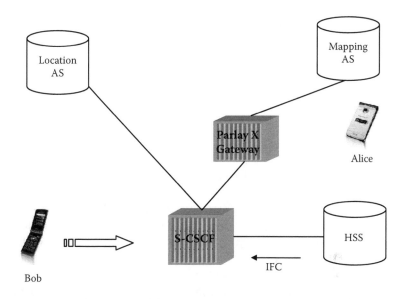

Figure 20.2   Example of configuration architecture for interactive applications.

Because the latency requirements are not as stringent for interactive services when compared to real-time services, many interactive services can be hosted external to an operator's network and connect through an external gateway such as a Parlay X Gateway.[4] A broader range of applications can be blended in this architecture as more Web-services-based applications are expected to be available versus SIP-based applications due to a broader developer's base. Although it is possible to have a real-time application using a Web-services-based application, they are not expected to be plentiful as Web services applications are considered "heavy-protocol" applications (higher latency) and thus tend to require a more relaxed latency requirement.

20.1.3 Background

An application running in the background can have a wide range of latency requirements but is typically characterized as reacting to some event such as an incoming call, a roaming situation, some form of a notification, or a particular time-of-day/day-of-week (TOD/DOW).

An application running in the background can be combined with a real-time or interactive application as IMS can support several sessions existing at one time. Continuing our previous example, Bob has subscribed to a weather service that provides his local weather forecast. This is a useful service for Bob since he travels often. However, he does not always know the zip (postal) code that he is traveling through. Thus, a simple example would have a weather application blended with a location application to deliver the local weather in a text or MMS message early in the day. Since it is immaterial to Bob whether the message arrives at 6:30 a.m. (local time) or 6:38 a.m., the latency requirements are not stringent. In this example, Bob is not taking any action to send his location to his weather application but instead the weather application is initiating the query in the background to obtain his current location.

20.1.4 Enablers

The real-time, interactive, and background classifications are really intended to show how different types of applications can be blended together to form new services. Blending is not limited to just applications, but can be extended to enablers which by definition can be used for multiple related or unrelated applications. The most commonly known enablers would be Presence or Location. By themselves, Presence or Location are of limited value since they must be placed in context with a service. However, blending their results with an application that can put the information into a useable form by the end-user is valuable.[5] In addition to Presence and Location, OMA has specified other enablers to be accessed by an application or in support of an application through a well-defined interface. Table 20.1 provides the list of currently defined OMA-approved enabler releases.

The key idea that should be reiterated here is the flexibility in bringing together different types (or classes) of applications along with well-defined network enablers to create new and rich end-user services. Gone are the limitations of non-IP-based networks where applications were limited in how they could be blended and were limited to a single protocol interface. IMS allows for blending of different applications in flexible ways plus through such defined elements as the service broker and the OSA Gateway, applications implemented using non-SIP protocol interfaces can be supported.

20.2 Service Delivery Framework

In the previous section we discussed blending within and between different classes of applications as well as blending with enablers such as Presence or Location. In Chapter 4 and Chapter 6 we looked at different methods to blend applications together as well as mechanisms to go beyond just SIP-based applications. However, one needs to look at the entire service ecosystem including service development, implementation, provisioning, billing, operations and maintenance, and even end-of-life scenarios. These different steps along a service's life cycle have tended to be very specialized to a particular operator, which has translated into a very slow and expensive process for any operator to deploy a service. This is an industry issue not limited to just IMS-based networks. The Information Technology (IT) community has recognized this concern for several years now and has been working

Table 20.1   OMA-approved enablers

Browsing
Charging
Client Side Content Screening Framework
Data Synchronization
Device Management
Digital Rights Management
Domain Name System
Download Over The Air
E-mail Notification
Firmware Update Management Object
Instant Messaging and Presence Service
IP Multimedia Subsystem[9]
Multimedia Messaging Service
Online Certificate Status Protocol Mobile Profile
Presence SIMPLE
Push-to-Talk over Cellular
Secure User Plane for Location
SyncML Common Specification
Standard Transcoding Interface
User Agent Profile
vObject Minimum Interoperability Profile
Web Services[6]
Web Services Network Identity
XML Document Management

toward developing a standards-based framework to manage the rapid time-to-market needs to support the back office requirements for new services. Their focus has been on the traditional BSS/OSS systems and not necessarily the network aspects of a service. This standards-based framework has been coined in the industry as a Service Oriented Architecture (SOA) and is based on Web services technologies.[6] The SOA effort has led to the development of some basic principles for any SOA, namely the concept of Functions as services such as Governance, Loose Coupling, Late Binding, Discovery, and Openness.

With the general acceptance of SOA as a means to support rapid production in an IT environment, more traditional network equipment providers (NEPs) along with certain traditional IT vendors started looking at ways to adopt SOA principles to the traditional network. This effort became known as Service Delivery Platform (SDP). Unfortunately and most confusingly for anyone trying to understand the topic, so many definitions were espoused for an SDP (primarily based upon what was synergistic with the product strategy of the company espousing the definition) that the term SDP has started to move out of favor. Also, as more study is done it is clear that all the components necessary to accomplish an SDP cannot be accomplished by a single "platform"[7] but instead it will require multiple platforms, each covering a particular area to provide a framework for supporting the total ecosystem. Thus, the new term Service Delivery Framework (SDF) is being used to acknowledge the need to incorporate both IT elements and network elements working together in a cooperative arrangement to address the holistic requirements in rolling out a new service. It should be mentioned that the SDF is not limited to a pure IMS (SIP AS) environment but can also incorporate other environments such as legacy IN (e.g., CAMEL, AIN, etc.) and Internet (e.g., XMPP, etc.).

Figure 20.3 shows one representation of the many different touch points that an SDF needs to make in order to successfully launch a new service. This representation

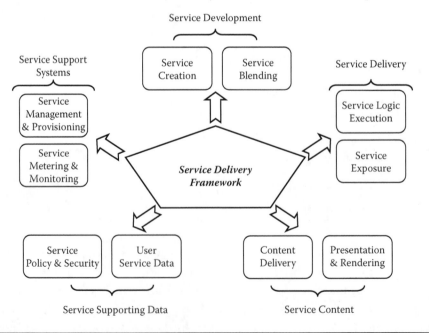

Figure 20.3 Major functional elements of an SDF.

shows how an SDF must cross both the network and IT domains. We will examine each of these functions in more detail.

20.2.1 Service Development

20.2.1.1 Service Creation Environment

The Service Creation Environment (SCE) provides a development environment for service creators to design, build, and test their new application. Many of the same concepts for earlier IN-based SCEs are still applicable. Functions or actions that comprise various network capabilities are broken down and defined as independent, reusable building blocks that can be arranged with other building blocks to build new applications as shown in Figure 20.4.

Operators looking to maximize the suite of services offered to their customers will want to allow the third-party developer access to their SCE, as well as to their internal developers. By providing access to their service development tools, plus a complete testing environment, service providers can more easily partner with the large Internet developers' community. The final product developed in the SCE (for both internal and third-party applications) could be hosted by the service provider on their SLEE (see next section) or the third-party developer

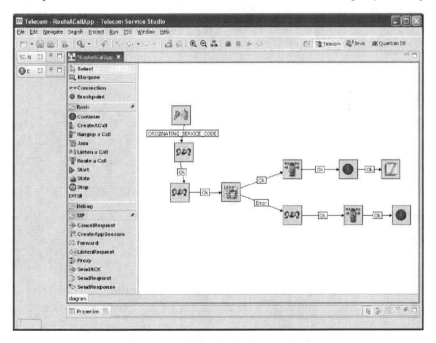

Figure 20.4   Service building blocks for an SCE. (From jNetX. Used with permission.)

could choose to host it themselves external to an operator's network and expose the application to that operator's customers through the operator's Parlay/Parlay X/Web Services Gateway.

20.2.1.2 Service Blending

Service Blending is taking existing individual applications and blending them together to create a new and richer application. By reusing individual applications that can be completely independent of other applications, the service provider can launch a new service that is quicker to market by leveraging already deployed functions, which in turn reduces the deployment cost of the new blended service.

In the IT domain, service blending is achieved by combining various services known as *business processes* to create new blended services (in SOA terminology, this is known as *service orchestration*). As mentioned earlier, these business processes are based on Web services using languages such as BPEL or BizTalk, which are flow/process-driven approaches. In the network domain, service blending can occur in multiple places and in multiple layers (Figure 20.5). In this domain, the new blended service is very event-driven, as well as data driven. For our purposes, the primary elements where blending will occur within an IMS network will be

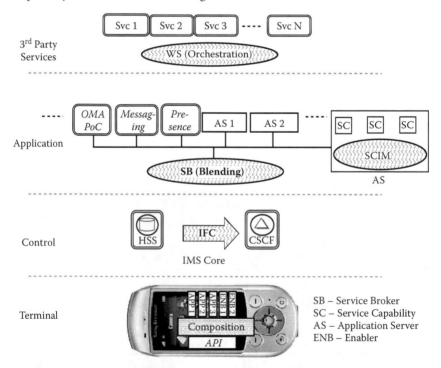

Figure 20.5 Layers where blending can occur.

through the service broker and the S-CSCF using iFC (these two elements were discussed in detail in Chapter 6). As shown in Figure 20.5, applications blending can also occur in the terminal device or in an application server; however, these tend to employ vendor proprietary solutions to bring about the blending action. The exception would be an application server that is based on existing standards such as JSR 289 (SIP Servlet) or a JSR 240 (JAIN SLEE).

20.2.2 Service Delivery

20.2.2.1 Service Logic Execution Environment

The Service Logic Execution Environment (SLEE) hosts the service logic for the applications that a service provider offers. In the IT domain, the service logic is commonly referred to as business logic execution, which supports the service provider's BSS/OSS systems. In the network domain, it is typically the service logic that is invoked (or executed) from an application server or service control point (SCP).

A SLEE container allows a programmer to focus essentially on the business logic of the application, letting the container to take care of issues such as protocol handling, high-availability cluster replication, application life-cycle management, and others. This simplifies the application development effort and reduces the time to development. The SLEE typically supports multiple technology environments. Many of these environments are based upon legacy (pre-IMS) IN technologies or vendor proprietary solutions. Two standards-based Java solutions have been specified. Of particular interest are the converged containers specified by JSR 289 which provides a convergent SIP/HTTP Servlet solution and the one specified by JSR 240 (JAIN SLEE) which includes a protocol agnostic container that can execute services over any network protocol.

20.2.2.2 Service Exposure

For applications running on a SLEE, they must have some mechanism to gain access to an operator's network in order to be accessed by the end-user. Operators must expose their networks through some well-known API. For exposure to third-party service providers, we examined in Chapter 6 the use of a Parlay or a Parlay X/Web Services APIs. Internal to an operator's network, we discussed the use of SIP interfaces as defined by Third Generation Partnership Project (3GPP™) architecture although an operator can use other interfaces internal to their network such as Web services or legacy IN protocols such as CAMEL, INAP, WIN, or AIN.

The IT Web services development community has adapted the Parlay X specifications as well for exposure to their Web services/SOA environment. Parlay X provides a much higher abstraction for Web services developers that encapsulate network functions or events. Such level of abstraction highly simplifies the invocation of network capabilities but at a cost of having a limited set of capabilities due to the fact that with such abstraction/simplification, the protocol details are not available anymore. Hence, the usage of Parlay X will remain limited for delivering some specific categories of applications that do not require complex network logic.

20.2.3 Service Support Systems

20.2.3.1 Service Management and Provisioning

One of the key beneficial claims of IMS is the ability to leverage the creativity of the Internet development community in order to greatly expand an operator's service portfolio in a rapid time-to-market manner. However, implementing a rich service creation and service blending environment to develop a plethora of services and providing a robust service execution or service exposure environment to run these many services will all be for naught if an operator cannot efficiently and correctly manage these services. This is the role of the Operational Support Systems (OSSs).

After a new service is ready to be offered, there must be means to provision, administer and maintain that service into the operator's OSS so that the network is aware of its existence (sometimes referred to as *service discovery*). In today's networks, these steps are typically unique to each operator, and a large effort is expended for each service to be individually supported. In addition, the end user must be able to be provisioned so they can gain access to the service. It is expected that as more services are rolled out, operators will chose to adopt a customer self-provisioning model in order to manage costs and training necessary for customer support staff. The Tele-Management Forum (TMF) is addressing this issue as part of the SOA architecture which many carriers' IT departments are adopting for their OSS infrastructure.

20.2.3.2 Service Metering and Charging

Similar to Service Management and Provisioning, an operator cannot expect to gain any revenue from offering new services unless they can properly meter and then charge for a service. Similar to their work for OSS, the TMF is addressing the Billing Support Systems (BSSs) as part of their SOA architecture. Currently, like the OSS, the BSS tends to consist of customized interfaces on a per-operator basis with significant effort to add each new service. By adopting SOA principles, the goal is to have a standardized interface that can quickly support new services.

There are also opportunities for operators to gain new revenue streams through their OSS and BSS networks. Operators can provide billing and customer care services to third-party developers who do not have the infrastructure or staff to handle billing or customer care. In these cases, accurate metering is essential in order to provide accurate billing record for any conflict resolution issues as well as the ability to offer service level agreements (SLAs).

20.2.4 Service Supporting Data

20.2.4.1 Service Policy and Security

Any service design must take into account the potential for fraud and abuse. Appropriate data settings must be available in order to determine the parameters allowed for the invocation of a particular service by a particular customer. Thus, service designers must include security and policy control mechanisms into their design, while operators must always look at ensuring the integrity of their network. Fraud can take many forms, including false or stolen identities used to obtain information or services or mechanisms to be assigned a service without a billing connection. Security procedures can be done using SOA governance in the IT domain and authentication/authorization information stored in the HSS, application server or some other network provided element in the network domain. Additional standards coming from the Liberty Alliance organization will facilitate single sign-on (SSO) functionality to provide a better user experience for the end user who accesses multiple services.

Abuse can come in deliberate or nondeliberate ways. As an example, an operator may set up a policy to allow a certain service provider to send messages into their network at a rate of 10 transactions per second based upon prearranged business agreements. If that service provider attempts to abuse the maximum agreed limit, then certain policies are invoked to throttle the throughput rate. Likewise, the service provider could implement policies against the end user to prevent various types of abuse. In either scenario, both the carrier and service provider must guard against legitimate devices or servers experiencing some form of a software glitch (such as an insane process or infinite loop) that causes a large stream of constant messages to be sent out, which could overwhelm a network if not properly protected.

20.2.4.2 Service and User Data

Data associated with the execution of a service along with user-specific data (such as preferences, etc.) can be located in multiple elements spanning both the operator's and the service provider's networks. From the third-party service developer acting from the service provider's perspective, it is important that they are provided a single point to retrieve data from the operator's network (typically, this would be the OSA/Parlay X Gateway). This is required to simplify the complexity of having to know all the data reference points. An operator may need to consider the 3GPP Generic User Profile (GUP) standards to support federation between its multiple databases in order to present all the service and user data through a single reference point.[8] Alternatively, they may consider a centralized user data repository to provide a single reference point to access service and user data.

20.2.5 Service Content

20.2.5.1 Content Management and Delivery

Following the Internet paradigm, many of the expected new IMS services will include some form of content delivery in their service blend. This SDF function addresses ensuring the content is available for the service from a work flow perspective. This includes addressing items such as the following:

■ Service discovery
■ Media adaptation
■ Digital rights management
■ Content acquisition
■ Screening or access control
■ Content storage and archiving

20.2.5.2 Presentation and Rendering

When one considers the "three screen" market (mobile device, PC, IPTV) and the difference in the size of the display screens, one quickly realizes the consequences of ignoring this SDF function with the end user. To display a 2×2 in. picture or video meant for a mobile device on a 50 in. LCD display will lead to a poor user experience as well as trying to fit a display meant for a 50 in. IPTV on a mobile device. This portion of the SDF must render (or custom fit) the downloaded content to the device that the end user is currently using. This is typically done by the operator as a way to shield the service developer from having to develop their applications for a particular type of device.

As part of the presentation, operators may add additional information in the form of branding, watermarks, messages, or advertisement to the content. These displays for the user's device can be custom designed, based on information from other parts of the network such as presence and location, or as part of the user's personal preference portfolio.

Endnotes

1. *Capability* is another term that is used the mix and thus needs to be defined as well. It typically refers to an underlying technology needed to offer a particular service. Examples would be a UMTS/HSPA transport bearer of SIP signaling.

2. One might notice the similarities between this classification and the four classes defined for QoS in Chapter 1. In the above classification, we are putting forth the position that live streaming services have the same real-time constraints as conversational services. Nonlive streaming services would have a more relaxed latency requirement, but for simplicity sake, we are not making the distinction.

3. The 3GPP2 standards have defined a method to support simultaneous voice-data applications called EV-DV (EVolution-Data/Voice). However due to time-to-market pressures, the CDMA carriers opted to go with the EV-DO technology, which precluded the EV-DV path.

4. The reader is referred back to Chapter 6 for a discussion on these external gateways.

5. Seeing a lat/long coordinate which a location server would provide is of little value to the common end user. However, having those coordinates translated to a known address or map point provided by an application is of significant value.

6. The use of SOAP as a service protocol is recommended as it is an extensible message-based protocol based on an XML schema. SOAP-based services express their service contracts using WSDL.

7. The term *platform* implies a single system or element, which has led to a lot of confusion in understanding the scope of an SDP.

8. To be fair, it should be noted that the GUP standards have lost a lot of traction in the industry due to the complexity of this problem. Many are moving toward a Liberty Alliance approach, although it is limited in fully addressing the data federation issue. However, for this discussion it is fair to state that the principles of the problem that GUP attempts to address need to be considered.

9. One can argue whether IP Multimedia Subsystem and Web services are truly enablers or not. Certainly other standards purists would argue against calling them enablers. For our purposes, we will note that OMA has categorized them as enablers and let others argue the appropriate taxonomy.

Further Reading

OMA Release Program and Specification, www.openmobilealliance.org/release_program.

"jNetX Service Delivery Framework (SDF)," White Paper (winter 2007), author Matthieu Loreille, http://www.jnetx.com/index.php?id=139.

www.projectliberty.org/resource_center/specifications Liberty Alliance Project.

3GPP TS 23.240, "3GPP Generic User Profile—Architecture; Stage 2."

"Service Delivery Platform—Efficient Deployment of Services," White Paper, October 2006, http://www.ericsson.com/technology/whitepapers/3083_sdp_a.pdf.

"Services in the IMS Ecosystem," White Paper, February 2007, http://www.ericsson.com/technology/whitepapers/3109_Services_in_the_IMS_ecosystem_A.pdf.

Chapter 21

Consumer Market Use Cases

Voice is the preeminent application in the consumer market, and it is expected to remain that way for years to come. However, with the commoditization of the voice minute, network operators are looking at ways to supplement the declining voice revenue with data applications (or rich voice applications).[1] While ubiquitous (mass market) services geared toward the entire consumer market (e.g., think calling line ID or SMS) will still be offered, these types of services will not be deployed at any greater frequency than is currently offered today.[2] So, given the need to supplement declining voice revenues, along with the constraint of limited mass market services, this leads to the logical conclusion that operators must go after the numerous niche markets within the consumer segment versus targeting the entire consumer segment. As we have seen in earlier chapters, IMS provides an environment that allows services to roll out in a more rapid fashion, which in turn allows the operator or service provider to focus on different target segments. Some examples of these targeted consumer segments that we will explore further in this chapter are

- "Soccer moms"
- Gamers
- College students

In this chapter, several examples of blended services will be presented that could be geared toward the consumer market based on well-known functionality or services. These new service blends can take place by combining different applications or different services in combination with an enabler. The example services will be introduced here using a known marketing description format called "A Day in a Life" scenario. In this format, a real-life scenario is envisioned to describe how a new service would be used by a targeted market segment—such as a consumer segment user. In the next three sections, we have created three separate vignettes (or scenarios) to describe how different IMS services could be used in a "real-life" situation.

21.1 Soccer Moms

Scenario

Jan is your typical busy single mother with an elementary-school-aged child, who is trying to juggle work and family obligations. Her son Ryan is a highly energetic nine-year-old who loves to play soccer and aspires to play in the World Cup tournament. Ryan plays for the Bulls, a youth soccer league team that practices every Tuesday and Thursday in the early evening hours ending at 6:00 pm. On practice days, Jan usually stops by the local grocery to pick up dinner for that evening's meal; however, this particular practice day is not going so smoothly. Jan arrives at the checkout counter later than she had planned due to a late departure from work. After frantically searching her purse she realizes she had taken her wallet out of her purse at the office to contribute to a colleague's retirement party and had left her wallet on top of her desk instead of putting it back in her purse. In a panic now, she breathes deeply and happens to notice that her grocery store offers a mobile payment option. Breathing again, she keys in her PIN number on her mobile phone as she places it near the payment reader.[3] Grabbing her grocery sack, she dashes for her car while she pulls up on her mobile phone the contact list for her work group. She quickly checks for the availability of anyone who might still be in the office. She sees from Eric's availability that he is in a meeting. Knowing her colleagues well enough to know they would be gone from the office by now unless they were in a late meeting, she decides to contact Eric to see if he could put her wallet inside a drawer on her desk for safekeeping until she goes back to work in the morning. Since Eric is in a meeting, he is forwarding all his voice calls to his voice mailbox. However, he is still receiving instant messages. Quickly, Jan types Eric an IM asking him to put her wallet inside her desk as she walks out of the grocery store. By the time she finishes loading her groceries in the car, Eric replies that he put her wallet in her bottom drawer and even included a picture of her wallet in that drawer.

Feeling better that her wallet is taken care of, Jan starts to leave the parking lot when she notices that her car is not driving correctly. She pulls back into a parking space to examine her car and sees that she has a flat tire. After a few choice words, she looks at her watch and realizes she needs to be leaving the parking lot now in order to pick up Ryan on time. Knowing this will not happen, she pulls up the contact list of parents on Ryan's soccer team to see if she could get one of them to pick up Ryan for her. Since she knows many of the parents are in carpools to pick-up and drop-off their children, she needs to determine which parents are actually at the practice field.

For situations just like this, the soccer team is part of a friend-finding group, so Jan sends off a quick location query to determine who is at the practice field. A map of the area around the practice field appears on her mobile phone's screen to show two parents are at the practice field now. One of those parents, Lisa, happens to live in her neighborhood, so Jan quickly calls her to ask if she could take Ryan home with her. After her call to Lisa, Jan sends a text message to Ryan's coach about the change of plans for picking up Ryan. And with Ryan taken care of, Jan can call her auto club to come change the tire.

21.1.1 Scenario Applications

In this vignette, several applications were invoked either individually or through a blended scenario in order to provide a service to the end-user. These applications were as follows:

■ Network-based address book (obtaining the location information of an individual group member will be explored in Chapter 23).
■ Network-based address book accessing a presence and availability server.
■ An IMS Messaging platform that allows for messages to be sent (and in this case, replied to) with an attachment.
■ Simple address book lookup for team coach and auto club. (Since the mobile phone's address book is synchronized with the network-based address book, this is a simple client address book lookup which provides no new functionality and thus holds no interest for further exploration.)

21.1.2 Call Flows and Network Layout

In our opening vignette, we show several applications involving the transfer of data (data-centric applications) that are used to enrich the voice experience in an IMS network. Presence is considered a key enabler that an IMS network can blend with other applications to enrich the end-user experience. A network address book is viewed as both an application and an enabler because it can function as both a stand-alone end-user service, and also it can be blended with other applications for a richer experience.

21.1.2.1 Network Address Book

Mobile phones are sold with an address book built into their product. In addition, for GSM devices, the SIM provides another mechanism to provide address book contact storage. The SIM-based address book has the advantage of being portable as a user switches or upgrades between devices. The device-based phone typically offers a more robust address book at the expense of portability. Neither has any advantage if the phone gets lost or stolen, nor is there any ease in sharing among multiple devices (such as a converged network scenario).

A network address book (NAB) offers the ability to transfer the address book content as the user changes devices, and it can provide for synchronization of content between a user's suite of devices (e.g., mobile, PC, IPTV, etc.). There are several other benefits to having a NAB. Among them are the following:

- If the customer's device is lost, their information can be easily restored to their new device without the need to reenter it themselves.
- If the customer's device is stolen, the network server can "synchronize" the stolen devices address book with null fields, thus wiping out the customer's contact information on that particular device to provide a level of privacy protection to the customer's contacts.
- Having a network-based address book allows a customer's contact information to be readily shared across different devices that the customer may be using (this would be a converged application scenario).

The NAB can be considered as a service, an application, and an enabler: a service in the sense that it provides a service directly to the end user (e.g., associates contact information); an application in the sense it can be a building block with other applications or enablers to form a rich service; and, finally, an enabler in the sense it is a data source into other applications needing a user's set of contact information. In the next section, we will look at the NAB being blended with the Presence enabler to form the rich service Presence-Enabled Address Book.

Address book standardization has occurred in the SIM area, but many operators view the SIM as falling short in offering a truly robust service to their customers. OMA is addressing this area through a new enabler (still under development at the time of this writing) called Converged Address Book (CAB). Among the features it will provide include the following:

- Common means to access from either an end user (subscriber) or another application
- Synchronization of contact information to the end user's device

- Manage the end user's own contact information for automatic distribution purposes
- Provides a search function for others to query in order to locate information about available users

This last item (search function) is similar to the XDMS search proxy that was discussed in Chapter 17. This can lead one to speculate that the ultimate end target would be to make the CAB align with the XDMS architecture (potentially as a service-specific XDMS that interworks or integrates into the shared XDMS). Since an XDMS stores records, it is ideally suited to host an address book's contact data. It remains to be seen what the final architecture will look like in its approved form. Figure 21.1 shows one example of what an XDMS-based solution would look like.

Other mechanisms have been defined for a network-based address book. OMA Device Synchronization (DS) is a widely supported standard to allow for client-server synchronization. OMA DS is based on the work of the SyncML Initiative, which merged into OMA at its founding, and the specification is also commonly referred to as SyncML.

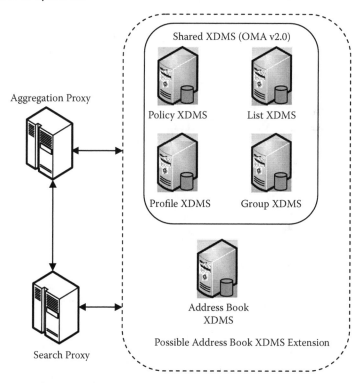

Figure 21.1   Possible XDMS-based address book.

21.1.2.2 Presence-Enabled Address Book

The Presence-Enabled Address Book (PEAB) is considered by many as a linchpin service for the deployment of a Presence and Group List Management Server. It is also viewed as a potential universal service that crosses the consumer, enterprise, and the converged markets. In fact, we will be exploring a variation of the PEAB as adapted for a converged market scenario in Chapter 23. The PEAB takes the concept of your standard mobile phone's address book and adds information about the presence status of a contact much like how the contact list for an instant messaging service identifies the presence status of the contact. It can also go beyond just identifying whether the contact is on or off the network. Indicators can be provided to show the availability or mood and the means by which the contact desires to be reached. Contacts can also be segmented by specific groups identified by the customer. Figure 21.2 shows an example mock-up of what the user interface may look like on a mobile device. One approach may be to have tabs to separate different groups of contacts (in our example, Jan has a work contact list and her son's soccer team contact list).

A PEAB would typically be a network-based address book versus a terminal or device-based address book, in part due to the lower traffic burden it would place on the Mobile Network Operator (MNO) air interface. Information from the network server would then be synchronized with the terminal's client address book for fur-

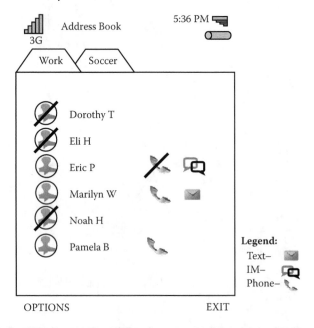

Figure 21.2   Example mock-up of PEAB screen.

ther network efficiencies (this limits the need to send network-based information on every address book look-up) and service response time requirements.

Figure 21.3 represents an implementation of a reference architecture for the deployment of a PEAB. The reader is referred back to Chapters 16 and 17 for the detailed discussion on Presence and Group List Management, respectively. In this figure, we are assuming the NAB is implemented as an application-specific XDMS.[4]

The NAB fulfills the role of the watcher in the OMA and Third Generation Partnership Project (3GPP™) Presence architecture. In this example, it would subscribe to the Presence Server to receive the presence status of each individual in Jan's work contact list and youth soccer team contact list. Figure 21.4 shows the NAB sending a request (SUBSCRIBE) to the Presence Server to receive the presence status for the contacts in Jan's work list and youth soccer list. Note that a single SUBSCRIBE message could be used to make the request for status information; however, for this example we will assume that the subscribed service level for each contact list is different. In this case, the presence status for each of Jan's work contact list is updated every hour on the hour (here we assume Jan's work would be paying for this service and the need for current status information), whereas Jan's youth soccer team is only updated on a demand basis (i.e., user requested) with a permission acknowledgement required from each soccer team parent (here we assume Jan is paying for this service herself, and the need for presence status information is sporadic, thus a lower service quality is sufficient). For both lists, each address book contact would be the presentity as described in Chapter 16.

Jan, as well as her contacts, can provide an update of their presence and availability status on a manual basis, in addition to the presence server gleaning the update from network-based information. In our example, we saw that Eric indicated that his availability status was "In Meeting." We saw in Chapter 16 the

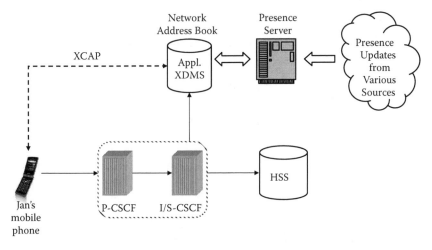

Figure 21.3   **Reference architecture for Presence-Enabled Address Book.**

different Availability values as well as Moods that had been defined. It will be up to the service provider and their vendors to determine which, if not all, of these values or even new carrier-specific values will be offered to their customers. It is expected that, depending on how the service is targeted, not all the values will apply, and thus not all will be offered for that particular service. Figure 21.5 shows how Eric provided an update of his availability status.

21.1.2.3 Instant Messaging

Instant Messaging (IM) has been a ubiquitous application on the PC, and IM clients can be found on most mobile devices as well. Currently, mobile IM is implemented

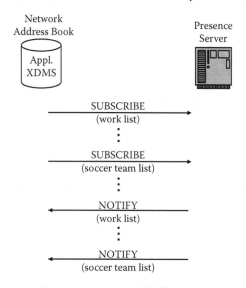

Figure 21.4 Presence status request and response.

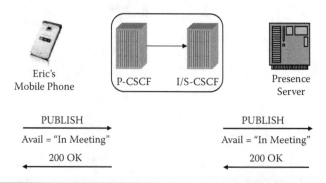

Figure 21.5 Updating of presentity's presence and availability information.

using the OMA Instant Messaging and Presence Services (IMPSs), more commonly referred to as the Wireless Village.[5] While IMPS meets the needs for a pure IM play, it has limitation with regards to being blended with other applications and working within an IMS network. Hence, OMA has capped the work on IMPS at Release 1.3 in favor of OMA SIMPLE (SIP Instant Messaging and Presence Leveraging Extensions). As its name implies, SIMPLE is a SIP-based protocol, which allows it to be blended and to interwork cleanly with other IMS applications.

As we saw in our PEAB analysis, the presence status of the target contact (in this example, this would be Eric) was obtained through the use of the SUBSCRIBE/ NOTIFY exchange (see Figure 21.4). This would be analogous to what occurs when one opens up their buddy list from their IM application on the PC and sees the presence status of one's buddies.

When Jan actually sends the IM to Eric, her device's User Agent Client (UAC) would populate what she writes into a SIP MESSAGE for delivery to Eric's device. To support a simple text message like what Jan sent would require each device to support the Content-Type: plain/text format. Other formats can be supported which would allow one to include graphics or video. It is this ability to support different formats that will permit the convergence of 2G messaging (i.e., SMS, MMS, IM) under a single message transport mechanism (the proverbial "a message is a message"). This will also allow for message interworking between the mobile and IP world since both would support the SIP MESSAGE method.

An example of the MESSAGE method is shown in Figure 21.6.

The message flow for sending the IM, including the check for the presence status of the targeted user, is shown in Figure 21.7.

21.1.2.4 Rich Communication Suite

In 2008, the GSMA assumed the responsibilities for a mobile operator and vendor-led effort that is designed to facilitate the adoption of IMS-based services by providing services that are interoperable across different vendor platforms and across different operator networks. The Rich Communication Suite (RCS) effort is not intended to define new standards, but instead it specifies a core set of features that

```
MESSAGE sip:jan@att_ims.net SIP/2.0
Via: SIP/2.0/UDP [5550::aaa:bbb:ccc:ddd]:1353;branch=z9hG4bKnashds6
Max-Forwards: 50
To: <sip:eric@att_ims.net>
From: <sip:jan@att_ims.net>;tag=31413
Call-ID: b89rjhnedlrfjflslj40a111
CSeq: 11 MESSAGE
Subject:  It's Jan. I need help.
Content-Type: text/plain
Content-Length: (...)

...Body of message...
```

Figure 21.6 Example SIP MESSAGE method.

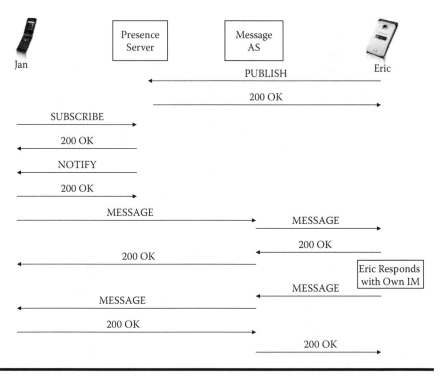

Figure 21.7   Message flow for Instant Messaging.

all parties will implement in a manner to ensure interoperability making use of existing standards. These industry-led efforts are important to ensure all parties involved interpret the standards alike.

RCS is expected to be a multiphase effort looking at different feature sets. The first phase will focus on three core feature sets:

■ Enhanced Phonebook: This feature set would include interaction with a network address book or other network-based directories (e.g., yellow pages) or serviced specific directory (e.g., game server). It would also support the presence status of the contacts to be displayed plus information on how the contact wishes to be communicated with.

■ Enhanced Messaging: This feature set allows the user to view all communication types (e.g., voice call, SMS, MMS, IM, etc.) in a conversational form similar to how a chat session is presented today, which shows the past (recent) conversation history.

■ Enriched Call: This feature set allows for the sharing of multimedia content during a voice call. It provides an indication of what multimedia sharing is possible at that particular instance in order to provide a friendlier user experience. An example would be if one user is in 2G coverage and thus unable to receive a video stream.

Much of the RCS effort has been focused on developing the mobile device client to support these three feature sets. However, there is recognition of the need to present the proper network interface to support a particular feature such as a NAB or to obtain presence information. Thus, this effort is also addressing interworking between client vendors and network platform vendors.

RCS is envisioned to address several of the examples shown in this chapter once the move to implementation is made. RCS will not address the user interface experience—this will be left up to the client vendor and the operator. Thus, in our first vignette, we examined the NAB service and the PEAB service, which would fall under the Enhanced Phonebook feature set of RCS. We will examine in later vignettes service examples falling under the other RCS feature sets—Enhanced Messaging and Enriched Call. Although we will not call out a direct reference to a service being implemented using RCS, it should be noted that RCS is an option to implement that service which is gaining industry momentum.

21.2 Gamers

Scenario

Alice has been looking forward to getting all dressed up and having an evening out with her husband, Bob, without any children present. She has obtained some front row mezzanine-level seats at the local opera house to attend a performance of Verdi's *Aida*, a performance she has always wanted to see. Unfortunately, for Bob to say that he doesn't enjoy opera would be to really understate his feelings about this musical form. However, Bob will get to share his misery with his fellow co-worker Bill, as Bill's wife Susan is friends with Alice, and they purchased tickets with seats together.

Leaving work the day of the opera, Bob quickly sends Alice a voice-activated IM saying he is leaving work now and heading home. Alice quickly responds with an IM to let him know that she stopped by their son's teacher's classroom to help her do some preparation work for the class and to ask him to pick up a pizza for the children.

Bob does a quick Yellowpages.com™ look-up[6] on his mobile phone for the number of their local pizza restaurant. He selects the number to call the restaurant and places his order. As he is about to pass the pizza store, Bob checks the status of this order and sees that it is ready for pickup.[7] He pulls into the local shopping mall to pick up his order and arrives home shortly after Alice. After quickly changing their clothes, they head off to the opera house to meet Bill and Susan.

The opera performance goes just as expected. Alice and Susan sit together thoroughly enjoying the performance while Bob and Bill sits on their outside thinking root canal work would be more enjoyable than this. Just before intermission, Bob takes out his mobile phone and clicks on the *aGames* icon.[8] He got introduced to Halo®3[9] a couple of months ago and is starting to get addicted to playing it. Wanting to see if any of his playing buddies are on-line, he pulls up his Halo®3 contact group in his Presence-enabled address book. He sees Tim, Peter, and John are on the network but the rest of his playing buddies are not present. Since Halo®3 is a "first-person shooter" type of game, it requires having a high-speed connection (e.g., 3G) in order to properly play. Bob does a quick request update on his gaming buddies to determine whether they are in 3G coverage.[10] The reply comes back indicating only Tim and John are in coverage. During intermission, Bob sends out a quick invitation to both Tim and John to request them to join him and Bill for a quick game of Halo®3. Since he has found a good game really requires six players he first asks Bill if he wants to join them in their game. Bill jumps at the opportunity so Bob sets up a temporary invitation for Bill to join his buddy list. He also puts a watch on Peter, hoping Peter would become present on a 3G network. Since he is still short of players, he puts out a general invitation on his gaming server for players looking for a pick-up match in 15 minutes. Since he prefers that Peter join them instead of an unknown player, he sets an indicator that if Peter should become available and want to play, that he would get priority over people who are not on his buddy list. Just as intermission ends, Bob walks over to Bill to inform him about who will be joining the game, then gives him a quick nod and says this should be a lot of fun.

As the third act begins, Bob checks the player status and sees that Peter has become available to play but does not accept the invitation. In his window screen he sees that he has a message, which he quickly checks. It is an IM from Peter with a picture attached saying he is at his son's basketball game and he cannot join them. The picture shows Peter's son Daniel dribbling the ball down the court.

Bob starts the game after a quick nod to Bill and taking confidence in seeing the wives mesmerized watching the opera. Bob, Bill, and Tim are on one team and John and the two pick-up players are on the other team. Bob's team does pretty good catching John's team in a couple of good ambushes. Since John works with Bob and Bill and they know him well, they have fun teasing him about his team's blunders as Bob sends obnoxious text messages to John. John gets pretty frustrated with losing so badly until Bill makes a mistake, which John's team immediately pounces upon and sends Bob's team fleeing down a strange corridor. Excited with this turn of events and wanting to do some trash talk back, instead of sending a text message back, John decides to send back some verbal abuse to all of Bob's team to the even the score using the Push-to-Talk (PTT) key at the bottom of the playing screen. With a jolt, Alice and Susan hear in stereo "Run away cowards" along with all the people around their seats. With horror both Bob and Bill realize they have both forgotten to turn off the voice option on the game. Well, needless to say, Alice and Susan are not very happy and are giving their respective husbands the death glare. And needless to say, the game ends quite quickly.

With the end of the opera approaching, Bob knows he and Bill are going to have to do some quick thinking to get themselves out of trouble. Opening up his mobile phone again, Bob clicks on his Personal Concierge Service icon to see if there is some place nice to take Alice and Susan after the opera. He finds the fondue restaurant that Alice has been mentioning that she wanted to try. He does a quick check on available reservations and finds one available in 45 minutes. With the reservation made, he does a quick survey of the restaurant's shopping mall, thinking he may need something more than just a nice restaurant to get out of trouble with Alice. He spots a florist shop that is opened late and quickly places an order for himself and forwards a quick note and the link to Bill with a suggestion that he does the same thing. Bob also sets the Status-Reply-Delivery trigger since the florist shop is part of the *Personal Business Concierge* network.

They arrive at the restaurant on time and are quickly seated. Bob does a quick glance at his mobile phone and sees the icon for the florist is green, indicating the order is being delivered now. Shortly after they order their dinner, the waiter arrives with two fresh bouquets of flowers. The wives are speechless at the extent of their husbands' thoughtfulness (or maybe it was groveling). Alice opens the note and sees it contains a picture from their last vacation (Bob had the foresight to deliver a photo from his phone's photo album with the text message for inclusion with the note card). After reading the note and seeing the picture, Alice leans over and pulls Bob toward her and gently whispers "OK, you're forgiven...but you owe me another night at the opera."

21.2.1 Scenario Applications

In the "Gamers" vignette, we introduced simultaneous multimedia applications and presence watcher functionality. The applications invoked in this vignette were as follows:

- Watcher feature on a member of the buddy list
- Directory look-up with availability status on purchase order
- Buddy list presence status with access network indicator (since the PEAB was previously discussed in Section 21.1, only the new features and functionality will be discussed here)
- IM with attached picture (previously discussed in Section 21.1)
- Push-to-Talk over Cellular (PoC)

21.2.2 Call Flows and Network Layout

With our gamers, we go to another level of details in examining how a Presence service would be implemented. We explore the role of the Watcher as well as the Presence user agent (PUA) and Presence network agent (PNA) as discussed earlier in Chapter 16. Different types of available Presence information are shown along with the use of a dedicated AS. We end our examination with a discussion of the PTT service.

21.2.2.1 Watcher List

Figure 21.8 provides the message flow diagram for obtaining the presence information of Bob's gaming buddy list. A couple of assumptions need to be made

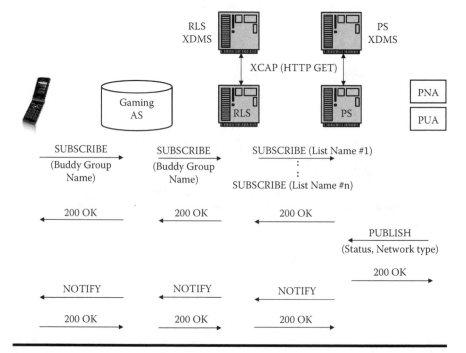

Figure 21.8   Watcher list update.

in explaining the figure. Our first assumption is around the Gaming Application Server (AS). A realistic implementation would have the Gaming AS serving as the primary watcher to the presence server, or, in other words, it would act as the watcher proxy for Bob. So here, Bob would not query the presence server directly but instead he must make the request through the Gaming AS. This is done for security purposes as the network will establish a known relationship with the Gaming AS. It is also done for commercial reasons as the Gaming Server provider could negotiate bulk rates for presence information as well as be able to access different operators' Presence Servers, as it would be expected that not all members of a buddy group list would be customers of the same carrier.

A second assumption is the Gaming service supports identification of players through a buddy list ID. This is a reasonable assumption as there will be incentives to either the player or the game service provider to reduce the number and size of those request messages when sent over a carrier's network. In our vignette, when Bob queries for the status of his playing buddies, he sends a SUBSCRIBE message containing the name of the buddy list, which is much more efficient (especially from an air interface perspective) than sending six (let's assume six since that was the number of players needed to make a game) individual SUBSCRIBE messages.

Our last assumption for the figure shows the separation of the Resource List Server (RLS) and the presence server into two physically separate entities (along with their respective XDMS). This was done to call out the exploder function of the RLS. Many of the Presence Servers on the market today are really a combination of these different functional entities defined by OMA; namely, one should expect to see a product that would be a combination of the Presence Server, PS XDMS, RLS, RLS XDMS, and the Shared XDMS. This integration of functional entities allows for certain performance improvements, simpler installation, and integration, as well as fewer complications to operate over a nonintegrated solution.

Referring back for Figure 21.8, Bob sends a SUBSCRIBE[11] message to his gaming service provider requesting the presence status of each of the players on his buddy list. The Gaming Server would process Bob's requests and send its own SUBSCRIBE message to the RLS with a group list ID. The RLS in turn needs to interpret the group list ID by finding out the particular member IDs of the group list. It does this by sending an XCAP query to the RLS XDMS with Bob's group list ID and getting a response listing the members of Bob's group (e.g., Tim, Peter, John, etc.) and their individual IDs (e.g., TEL URI).

The RLS in turn would send the presence server a SUBSCRIBE message for each member of Bob's group to obtain their individual presence status. The presence server would then send an XCAP message to the presence server XDMS in order to obtain the permissions/authorization status of the requester (Watcher or, in this case, Gaming Server) to get the presence status information on the members of Bob's group. In this example, we will assume that each member of Bob's gaming buddy list has given their preapproved permission (such that the presence server does not need to go out each time asking individual permission when a request is made) to have their status queried for the purpose of playing on the game server.

The presence server would then obtain the status of each member directly through a Presence User Agent (PUA) or indirectly through a Presence Network Agent (PNA). In our example here, the most likely case is the presence server will obtain the status information directly from the terminal (through the PUA in the terminal's client). This would be the case if each of Bob's gaming buddies had an IMS device.[12] A PNA would be required on terminal devices that could not publish their status or when newer information may be requested.

The presence server would send the NOTIFY message back to the RLS for each of the SUBSCRIBE messages it received. The RLS will gather these NOTIFY messages together for sending to their requester (or Watcher). Typically, the RLS will wait some finite period of time in order to gather as many (or all) of the NOTIFY responses as possible before sending its NOTIFY message in order to reduce the amount of traffic on the network. These time intervals are operator settings and will usually be dependent upon the needs of the particular application or a class of service level.[13] We will assume that the presence server had recent presence status information on each of Bob's buddies and was able to transmit status information on each of them with the class of service level time interval, and thus only a single NOTIFY message needs to be sent to the Gaming AS.

The Gaming AS would receive the NOTIFY message from the RLS. It would perform whatever needed task required for the execution of the game and send the status information of each member of Bob's buddy list down to Bob's terminal again, using a single NOTIFY message. Bob's terminal would receive the NOTIFY message and deliver it to the IMS client middleware. Depending upon the IMS client design, it may deliver the information to the address book application directly, which in turn would share this information with the gaming application, or it could deliver the information directly to both applications.

21.2.2.2 Directory Look-Up with Availability Status

A directory look-up from an IMS mobile device is primarily a Web-surfing exercise. IMS would play a limited role here except in its support to allow the mobile device to connect to various access networks. In our example, if we assume that Bob has a converged mobile device (Wi-Fi plus UMTS™/GSM),[14] then we can expand on our scenario as follows: When Bob places the order for the pizza as he is leaving work, we can assume he was still in or near his work building where he was still within the Wi-Fi coverage of his office building. As he approached the pizza restaurant, let's assume this was more of a rural area of his town where the 3G coverage had not been built out, and thus he received his order availability status over a 2G network. Finally, at the opera house (which we will assume is in the well-established downtown district), 3G coverage was available for him to make the restaurant reservation through his Personal Concierge Service. In all these cases, the same IMS network would be supporting Bob's services.

An application that is designed to send a status message will want to send it using the same method without having to worry about the network capability that the recipient is currently on. In addition, the developers of these specialized applications (in our case, a pizza order status notification and Personal Concierge Service notification) will most likely be developed and provided by third-party companies providing the service using an operator's network. Although the third-party service provides value to the operator and its customers, the operator must first and foremost protect its network from malicious or errant applications, plus it must have appropriate business measurement tools in place to ensure it is providing an agreed-upon Service Level Agreement (SLA). As discussed in Chapter 6 and Chapter 20, IMS does provide a mechanism to support this capability by providing an interface to an OSA/Parlay Gateway. In any real-world implementation, the OSA/Parlay Gateway will also incorporate a Parlay X Interface. Because of the software developers' community familiarity with Web Services development tools, it is expected that a large portion of the third-party application developers will write their applications using WSDL. Hence, the operator will deploy a Parlay X Gateway into its IMS network in order to handle the conversion between WSDL and SIP.

Since this application calls for a simple push notification of the order status, the application developer most likely will use a Parlay X short text message (SendSms)[15] to convey the status. The application server would send this message to the OSA/Parlay Gateway, which in turn would be converted into a SIP MESSAGE for delivery to the IMS network. The SIP MESSAGE would be propagated through the IMS network, which would have knowledge of the customer's serving network (e.g., on the macro network or a Wi-Fi network) as shown in Figure 21.9. The IMS would route the SIP MESSAGE through the Packet Data Gateway (PDG) serving the Wi-Fi network or through the SGSN serving the macro-cellular network.

21.2.2.3 Access Network Presence

When Bob sought to set up his multiplayer game session, the game itself required a certain player connection speed in order to maintain a certain quality of game playing to its players. With 3G networks still in their build-out phase, one cannot assume a particular user will always be under 3G coverage. With the emergence of converged devices, broadband access can be extended to those devices near a Wi-Fi hot spot if the user has access to that hot spot. The ability to determine a user's access network type ahead of attempting to invoke a service will provide a better user experience, as it can inform the user and the application of whether a service is even possible before an attempt is made to invoke the application.

3GPP defines two mechanisms through which presence information can be provided to the presence server: directly from a Presence User Agent or indirectly

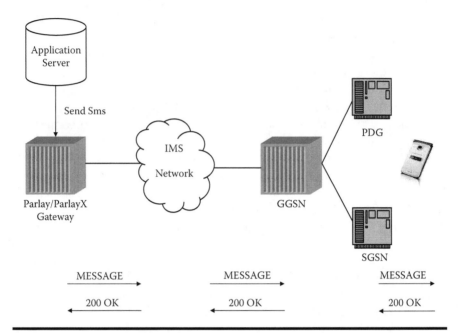

Figure 21.9   Push notification.

from a Presence Network Agent.[16] Expanding on the discussion in Section 21.2.2.1 and Figure 21.8, Figure 21.10 shows a broader view of these two entities.

In our vignette, it was assumed that all of Bob's playing buddies are using an IMS mobile or similar device (such as a PDA or laptop device) with 3G coverage. When each device publishes their presence status, they would include in that message the "P-Access-Network-Info" header to identify their network connectivity status. RFC 3455 identifies six access-type values that may be used:

- IEEE-802.11a
- IEEE-802.11b
- 3GPP-GERAN
- 3GPP-UTRAN-FDD
- 3GPP-UTRAN-TDD
- 3GPP-CDMA2000

So in our vignette, when Tim or John's device provided their presence status information, and assuming they were on a UMTS/HSDPA network, they would have sent a PUBLISH message similar to the following (Figure 21.11), which would identify how they were connected to the network.

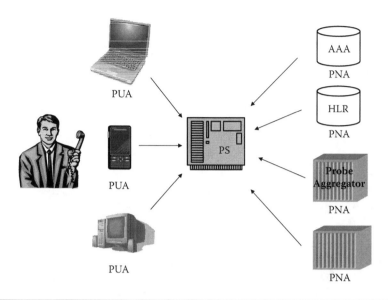

Figure 21.10   PUA and PNA sources.

```
PUBLISH sip:bob@att_ims.net SIP/2.0
Via: SIP/2.0/UDP
     [5555::aaa:bbb:ccc:ddd]:1357;comp=sigcomp;branch=z9hG4bKnashds7
Max-Forwards: 50
P-Access-Network-Info: 3GPP-UTRAN-TDD; utran-cell-id-3gpp=234151D0FCE11
Route: <sip:pcscf1.visited1.net:7531;lr;comp=sigcomp>,
       <sip:orig@scscf1.att_ims.net;lr>
P-Preferred-Identity: <sip:bob@att_ims.net>
Privacy: none
From: <sip:bob@att_ims.net>;tag=31415
To: <sip:tim@att_ims.net>
Call-ID: b89rjhnedlrfjflslj40a222
CSeq: 61 PUBLISH
Require: sec-agree
Proxy-Require: sec-agree
Security-Verify: ipsec-3gpp; q=0.1; alg=hmac-sha-1-96; spi-c=98765432;
     spi-s=87654321; port-c=8642; port-s=7531
Event: presence
Expires: 7200
Content-Type: application/pidf+xml
Content-Length: (...)
```

Figure 21.11   Example PUBLISH Message Identifying Access Network.

For a response from Peter who was not on a high speed (e.g., UMTS/HSDPA) broadband network, his P-Access-Network-Info header would indicate a value of "3GPP-GERAN," indicating that he was connected to a sub-broadband network (e.g., GPRS or EDGE connection).

21.2.2.4 Push-to-Talk over Cellular

Push-to-Talk over Cellular (PoC) is the telecommunications industry standardized version of the widely deployed iDEN™-based proprietary PTT service. Being IMS-based, the service is able to support multiple media types including both voice and video. Thus, similar setup call flows would be expected for a PTT and a Push-to-Video (PTV) call. As we have previously seen, by modifying the SDP parameters within the SIP message, one can obtain different types of media services. In Chapter 23, we will look at a PTV (sometimes called Push-to-Show) example.

We show the call flow to set up the PoC group session in Figure 21.12. For simplicity, we assume all group members are served off of the same PoC server, although this may not always be the case. Bob initiates the PoC session by sending an INVITE message to the PoC Server containing the group identity of his gaming buddies (Bill, Tim, John, and Peter). Most likely, an IMS gaming server will allow for multiple sessions (e.g., actual game session plus a PTT session) to run concurrently based on the subscriber's profile. Since Bob's profile contained only his four buddies with which to allow for "trash" talk, the two pick-up players were not invited to join the PoC session. The flow here illustrates a manual answer mode from Bill et al. versus an auto answer mode. This is indicated by the return of the 180 Ringing message.

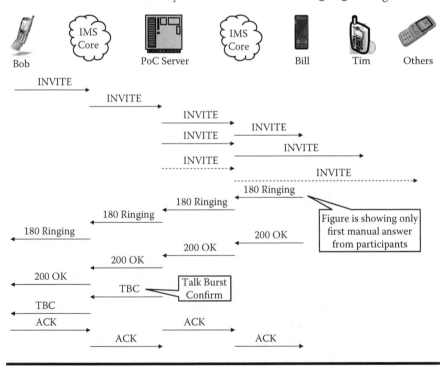

Figure 21.12 PoC call flow.

Since in our vignette Peter declined joining the game, his terminal device returns a 603 Decline message (although specific product implementations may return a different message; if the PoC server supports the SIP standard, it should be able to handle these variances). Finally, we see the standard acceptance/acknowledge with the 200 OK/ACK messages. The PoC standard includes additional messages to the "normal" setup message flow. Here the PoC server includes the Talk Burst Confirm (TBC) message indicating to Bob's device that the PoC session has been established.

21.3 College Students

Scenario

Heath, a 20-year-old athlete from Nashville attending Georgia Tech University in Atlanta, awakens at 7:45 a.m. to the music of his favorite band "Switchfoot" playing over his mobile phone and set using the *aCalendar*[17] application. His 7:45 a.m. wake-up alarm gives him just enough time to make it to his 8 a.m. history class. After rolling out of bed, Heath checks his mobile phone for any alerts or messages. He notices he has one alert message and one video mail message. Looking at his alert message, it is from his history professor sending an alert via the aCalendar to all students in the 8 a.m. class using the aCalendar Campus Alert.[18] Heath chooses the professor's alert, and a message pops on the screen along with the picture of the professor. The message reads **"For those who were sleeping in my class last time and missed my announcement, note that today's class will be cancelled. You can now sleep more comfortably in your bed than in one of my chairs☺"** Elated with the news, Heath flops back into bed.

Just as he is getting comfortable, he remembers he also has a message.[19] Rolling over, he grabs his mobile phone and presses the message retrieve icon button. It is a broadcast video voice mail from his soccer coach. **"This is your reminder notice to all of you who stayed up past curfew last night and forgot to set your alarms. The bus leaves at 8:30 a.m. sharp so make sure you are down at the locker room no later than 8:15. And by the way, make sure you bring your away-game uniform."** The message concludes with the coach holding up their yellow traveling uniform. Panic stricken at having forgotten about the game, Heath leaps out of his bed and begins to sort through two piles of clothes on his dorm floor; one clean clothes and one dirty clothes. Finding the correct soccer uniform, he remembers to throw in some clean street clothes, as this game is in his hometown and he plans to meet his girlfriend Lauren after the game.

Puffing as he arrives at the locker room at 8:20, he gets the "stare" from the coach who only says one thing to him, "Got the right uniform?" Heath breathes a sigh of relief as he pulls out his uniform to show the coach.

On the bus now, Heath opens up his phone to see that a new message has arrived. The message is from Lauren and reads "Meet you after the game; looking forward to seeing you. Lauren." A new picture of Lauren is also attached to the message.

Lauren has been on MySpace™ updating her page for the past hour. Knowing Heath is coming into town is exciting, as she has not seen her boyfriend in a couple of months. She posts "One year anniversary with Heath" on her MySpace page next to a picture she had uploaded from her camera phone from the last time Heath was in town. She then drags the link into a message she sends to a few friends: "Heath is playing tonight at the soccer stadium, want to come?"

Meanwhile, after a six-hour bus ride, Heath's soccer team arrives at the soccer stadium. Warm-ups quickly begin as the game is fast approaching. The stadium is beginning to fill up as some of Heath's hometown friends, including Lauren, begin to show up. Only one of Lauren's friends, Lindsey, is there with her, but she promised to keep the others involved remotely. She is talking to Paige on her Video aShare[20] phone about meeting up with her and a couple more friends after the game. Paige has put Lauren on the speaker phone so she can hear the other people with her. Lauren opens a video share session with her friend from the stands and points the camera at Heath who is warming up.

The game is closely matched, and Lauren and her friend enjoy the game until midway through the second half when Heath's team makes a run for the goal. It is at this point that Heath makes a tremendous leap to head the ball when he collides with the opposing team's goal keeper. The collision can be heard even over the roar of the crowd. The keeper lies motionless while Heath is writhing in pain, holding his knee. The medical team rushes onto the field where they are able to get the keeper back on his feet but not so with Heath. They whisk Heath off to an ambulance as the final seconds of the game tick away. "Hey coach, I'm going to ride in the ambulance with Heath," says Robert, the team's student trainer and Heath's best friend.

Lauren tries to call Heath on his mobile but cannot reach him, as all his belongings are still in the locker room. She goes to her address book and pulls up Robert's number, figuring he would be with Heath. After briefly chatting with Robert, she adds on Heath's parents to the call to let them know what has happened. "Is Heath alright?" his parents ask Robert. "His leg is pretty banged up. They won't know all the details until we get to the hospital and get it x-rayed, but it doesn't really look that bad. Do you want to talk to Heath?" Heath takes the phone from Robert. "Hey Dad, I'm O.K. I think I messed up my knee though. They're taking me to the hospital. I'll call you when I find out more." "O.K., do you need us to come down?" "No, Robert and coach are here. Bye."

After hanging up, Heath sends a short IM to Lauren: "I'm OK. Meet you at the restaurant after I get out of here." Seeing this, Robert quickly adds, "Yeah, in your dreams."

After the game a couple of players meet Lauren and her friend. "What's up, Lauren?" asks Pete, one of Heath's teammates. "I'm worried about Heath." "He's going to be fine. Let's go on to the restaurant he was going to meet you at and wait there." The five of them jump in Lauren's car and drove off to the restaurant.

Back at the hospital, Robert opens up his mobile phone and takes several pictures of Heath's leg as they are wheeling him into the hospital emergency room. He hides in the back of the room and messages Lauren, attaching the pictures: "Lauren, Heath won't make it to the restaurant tonight. I'll keep you posted."

While waiting for the x-ray room to open up, a weathered, rather rotund nurse steps up to Heath and gives him a gown to put on. "I'm not putting that thing on!" Heath says emphatically. "You will or I'll make you put it on!" the nurse answers back. Robert starts laughing and the nurse shoots him a piercing glance. "I'll put one on you, too, if you keep it up!" After a few minutes, the nurse comes back and yells "Heath." Heath raises his hand. "That's me." She waives at him from down a cold, white hallway "Come on back." Heath starts hobbling down the hall and his hospital gown opens up in the back just as Robert turns in his direction. He starts to laugh as more and more of Heath is exposed. "I gotta get this on film." Robert thinks to himself. He begins to tiptoe behind Heath down the hallway and whips out his mobile phone with Video aShare blazing. He goes to his phone's address book and finds Lauren's number.

Back in the restaurant where the players are eating dinner, Lauren's phone goes off. As she answers the phone, she hears Robert say, "Here's the man of your dreams" and she quickly lets out a loud laugh. "What's going on?" Pete asks, as Lauren holds up her phone to reveal the highlighted backside of the goal scorer hobbling down a well-lit white hospital hallway. "It looks like the man of my dreams won't be joining us tonight," smiles Lauren.

21.3.1 Scenario Applications

In the "College Students" vignette, we introduced calendar interaction, video voice mail, and peer-to-peer video share application. The applications invoked in this vignette were

- Community text message interaction
- Video voice mail
- Simultaneous voice call with one-way video stream
- Three-way conference call

21.3.2 Call Flows and Network Layout

Our previous applications mainly centered on data-type services working independently of a voice call. Here in this vignette, we will examine services that include two or more users engaged in a traditional voice conversation that is blended with a more data-centric application to provide a richer service experience to the end user.

21.3.2.1 Community Text Messaging

So far we have only shown application servers existing within the network operator's network. In this example, we show that other elements can occur in the third-party ecosystem (northbound of the Parlay/Parlay X Gateway). In Figure 21.13 we

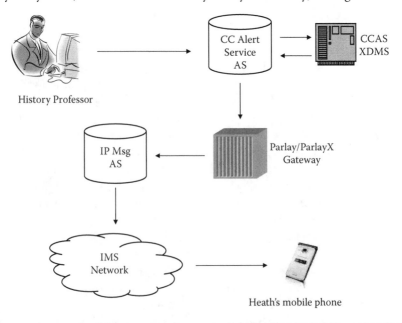

Figure 21.13 Campus calendar alert system architecture.

show an application-specific XDMS, namely the Campus Calendar Alert System (CCAS) application. Since the XDMS can also be used to manage data groups or lists, one of the groups contained with the CCAS XDMS would be all the students in Heath's 8 a.m. history class. So in our scenario, Heath's history professor could log into the CCAS using any Web browser interface and, after any application-specific authentication procedure, it would present to the professor a selection screen from which he could select the particular class and type in any specific message (in this case, a cancelled class reminder).

Receiving this request, the CCAS AS would query its associated XDMS using XCAP to retrieve the class contact information. Note that although XCAP is the defined communication mechanism between these two logical entities, vendors may choose to implement them as a single physical entity and use a more light-weight or proprietary protocol between these two elements in order to improve the transaction performance over the much heavier XCAP protocol.

The message returned from the CCAS XDMS would contain a list of URIs corresponding to each of the students in the 8 a.m. history class. These URIs would take the form of a TEL URI (e.g., TEL: 123-456-7890) if the student listed their mobile or SIP phone as a contact point or as a SIP URI (e.g., SIP: heath@gt.edu) if they listed an e-mail address as a contact point. The CCAS would take the address of record for each student, attach the text message from the history professor (in our example), and send individual messages out over a Web services interface to either an operator's Parlay X Gateway (for TEL URI addresses) or an e-mail gateway (for SIP URI addresses).

At the Parlay X Gateway, authentication of the message source is provided followed by the application of any Service Level Agreement (SLA) with the message source. The Web Services message is translated into a SIP MESSAGE at the Parlay X Gateway and forwarded to the IMS Messaging AS over an ISC Interface. In a transitional phase to an all-IMS network, and depending upon the capabilities of the Parlay X Gateway, the messages could also be forwarded to a legacy (non-IMS) Short Message Service Center (SMSC) using a SMPP[21] Interface for non-IMS recipient devices.

Finally, at the IMS Messaging AS, it would take each message it received and delivery it to the appropriate recipient. Figure 21.14 shows the message flow for delivery of the SIP MESSAGE from the IMS Messaging AS to the recipient's device.

The IMS Messaging AS forwards the MESSAGE request to the I-CSCF since it does not know the serving network of the recipient. In the MESSAGE it would place the address ID of the recipient in the P-Asserted-Identity field and would take the form:

```
P-Asserted-Identity: <sip:heath@gt.edu>,
<tel: +1-404-555-1234>
```

The S-CSCF of the recipient is determined through a user location query to the HSS as discussed in Chapter 4 and Chapter 14. The request is sent over the Cx

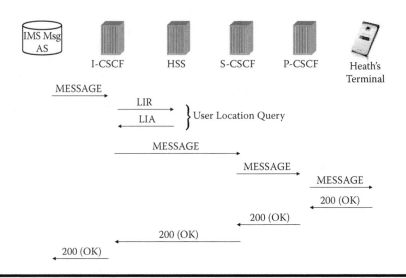

Figure 21.14   Message flow from IP messaging AS to terminal.

interface (DIAMETER) in a Location-Info-Request (LIR), and the response (address of S-CSCF) is contained in a Location-Info-Answer (LIA) message. The MESSAGE is then sent to Heath's S-CSCF. The S-CSCF has the registration information about Heath on which P-CSCF is serving his device. The S-CSCF forwards the MESSAGE to the appropriate P-CSCF which in turn delivers the MESSAGE to Heath's mobile device.

21.3.2.2 Video Voice Mail

The migration to IP continues to add credence to the marketing mantra "a message is a message." Whether a message is a store-and-forward text (SMS), multimedia (MMS), session-based (IM), or voice (voice mail), all can be delivered via a single mechanism versus the silo methods of non-IMS networks. Our previous examples have shown this in the traditional areas of data messaging (e.g., SMS, IM); here our example shows that a voice mail message (and in our specific example—a video voice mail) can be delivered in the same manner as other data messages. As a historical note, video voice mail under a 2G (non-IMS) network was typically delivered as an MMS message using standard MMS procedures. While this method enabled the service to be deployed, it required multiple network elements (SMSC, MMSC, IM Gateways) to be deployed to support the different messaging methods.

Figure 21.15 shows one instance of a deployment architecture for video voice mail where we show the voice mail platform loosely defined as consisting of three functional components: Voicemail Service Logic, Voicemail XDMS, and the IMS Messaging AS. The Voicemail Service Logic would cover the service logic as well as the traditional TUI (telephone user interface). The Voicemail XDMS would contain

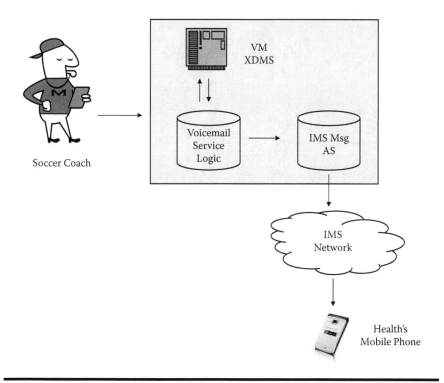

Figure 21.15 Video voice mail architecture.

the group distribution list for the broadcast video voice mail message. The IMS Messaging AS would act as the store and forward server for the video voice mail message and the platform to package the voice mail message for delivery into the IMS network. Figure 21.15 is drawn to show the IMS Messaging AS as a component of the voice mail platform. While most real-world vendor deployments will have the functional capabilities of the IMS Messaging AS integrated within their product, the figure is drawn to show that it could be separated from the voice mail platform itself and deployed as a stand-alone product. This would then give us the same architecture as shown in Figure 21.13 (minus the Parlay/Parlay X Gateway, of course). Vendors and operators could chose to have the IMS Messaging AS integrated with the rest of the voice mail platform for reasons such as scaling and capacity, different performance requirements, application diversity in case of major failure, and for operators if they hold onto legacy silo business models in order to allocate costs.

As would now be expected, the message flow for the video voice mail will be the same as for the community text messaging as shown in Figure 21.14. SIP MESSAGE would be used to the deliver the video voice mail. The difference between these two cases would be the MIME type identification carried in the SIP MESSAGE. In our voice mail example, the SIP MESSAGE would contain a MIME type specifying a video message such as video/3gpp.

21.3.2.3 Simultaneous Voice Call with One-Way Video Stream

When Lauren set up her Video aShare[22] call to Paige, she established a direct peer-to-peer session with Paige's phone. For this example, we will assume the current GSM implementation of the Video Share service which uses the existing circuit switch network for the voice path and the 3G data network to carry the video stream. This example will show how two terminal-based devices can establish a session between themselves without direct network knowledge of the actual service being invoked. Previous examples discussed (e.g., Section 21.2.2) have shown services being invoked using a network-based application server, which would provide the necessary mediation for two devices to communicate. In our example here, no such application server is present, and thus the two devices must perform the service establishment mediation between themselves. Figure 21.16 shows the architectural layout for the peer-to-peer video service.

Some may assume (incorrectly) that because the applications are contained in the terminal devices and negotiation is done independently of the network that there is no added value provided by the network nor is there a need for IMS. Many people new to IMS have a Vonage™ or Skype™ model in mind as they put forth the analogy of two PCs mediating a peer-to-peer session to establish a VoIP connection. What is missed in this analogy and a key value proposition that IMS provides in a peer-to-peer application is the necessity to address the mobility nature of the end device.[23] Whereas in our PC-PC example where the end points are fixed, IMS

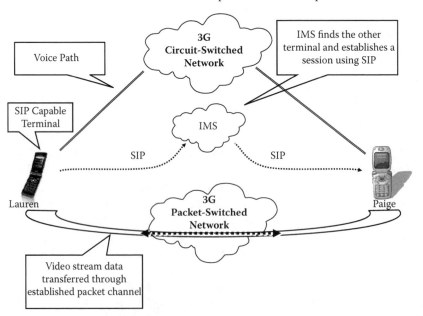

Figure 21.16   Peer-to-peer video share architecture.

supports the ability for the end points to be mobile and to be found through a query to the HSS.

From our vignette, Lauren calls Paige from the soccer field. When that call is placed, a subsequent message exchange is initiated between the two devices to determine the capabilities of the two devices. Figure 21.17 shows the message flow for discovering the capabilities of the far-end terminal device in order to set up a Video Share session. For simplicity sake, we assume standard registration and voice call completion procedures as discussed in Part I and Part II have already taken place.[24]

The OPTIONS exchange can be initiated by the device client independent of the user to understand the capability of the far-end device. This is done for user-friendliness reasons. In the transition to ubiquitous IMS and 3G coverage, there will be certain services that require certain network capabilities in order to run properly. In our specific Video Share example, a high speed (e.g., 3G or Wi-Fi) connection and an IMS device are required of the terminating party (a high-speed connection is also required for the originator).[25] It can be a deterrent to the user if she attempts to place the video share call and constantly receives a failure notice with no explanation. Human-factors studies have indicated that after just a few failed attempts, the user will stop attempting to invoke the service because she has been trained to expect failure. A smarter implementation of this service will first check to

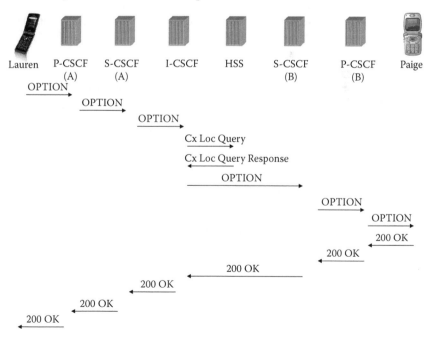

Figure 21.17   Discovery process using OPTIONS message.

determine if the service is even capable of succeeding (i.e., OPTIONS exchange) and then control the activation buttons accordingly. In our example, Lauren will see her Video Share "activation" button light up if the service knows a priori that it will succeed. If the OPTIONS exchange indicates that the service cannot work properly, then the "activation" button could be grayed out to give Lauren an indication that either the person she is calling does not have a device capable of receiving a video call or one of the two of them is in non-3G coverage.

Looking at Figure 21.17, we see Lauren's phone initiate sending the OPTIONS message to Paige's mobile phone using standard SIP procedures and the responding acknowledgment in the 200 OK message. Since the OPTIONS message is requesting the capabilities of the receiving device, negotiation does not occur as happens when an INVITE message is sent as part of an Offer/Answer model.[26] This is because the OPTIONS message exchange does not establish a session but only provides information on whether a session is capable of being established. So the key item of interest here is the SDP message body contained in the corresponding acknowledgment message (200 OK), which lays out the capabilities of the device—in our example here, to establish a video share session.

So an example of an OPTIONS message may take the form in Figure 21.18.

Although an SDP body may be included in the OPTIONS message, it may not be mandatory as no negotiation is taking place. The key field to note in our above message is the Accept-Contact, which carries the feature tag +g.3gpp.cs-voice. This is the feature tag used in GSM to inform the receiving end that the sender supports a normal circuit switch voice call as part of the Video Share service and asks for its capabilities in supporting a Video Share call.

Next, we examine the acknowledgment (200 OK) message, which lays out what Paige's mobile phone can support. For simplicity, we will focus on the SDP body and omit here the SIP message headers. We will just note that the Call-ID in the 200 OK message is the same as shown above in the OPTIONS message so Lauren's mobile device can correlate the response message with the proper sent message.

```
OPTIONS sip:paige@att_ims.net SIP/2.0
Via: SIP/2.0/UDP
      [5555::aaa:bbb:ccc:ddd]:1359;comp=sigcomp;branch=z9hG4bKnashds8
Max-Forwards: 50
P-Preferred-Identity: <sip:LAUREN@att_ims.net>
Accept: application/sdp
Aceept-Contact:*;+g.3gpp.cs-voice=TRUE
Allow: INVITE,ACK,CANCEL,BY,OPTIONS
From: <sip:lauren@att_ims.net>;tag=31415
To: <sip:paige@att_ims.net>
Call-ID: b89rjhnedlrfjflslj40a333
CSeq: 1 OPTIONS
Contact: <sip:lauren@att_ims.net>;+g.3gpp.cs-voice
Content-Type: application/sdp
Content-Length: 0
```

Figure 21.18 OPTIONS message layout.

```
Content-Type: application/sdp
Content-Length: 176

v=0
o=paige 28908155730 2890855731 IN IP4 host.att_ims.com
s={value can be empty}
c=IN IP4 host.att_ims.com
t=0 0
m=video 0 RTP/AVP 102 98
a=rtpmap:102 MP4V-ES/90000
a=rtpmap:98 H263-2000/90000
a=framerate:15
```

Figure 21.19   200 OK message SDP body.

For the SDP layout in Figure 21.19, we want to expand on the key field "media" (m=). The media field provides the video codecs supported by the receiving device (i.e., Paige's mobile phone). Recall from Chapter 12, Section 12.1.3 the layout of the media field:

```
m=<media> <port> <proto> <fmt>
```

Here the media is obviously video. The port is set equal to "0" because we are not trying to set up an active stream which would require a port. Our transport protocol will be RTP/AVP. Finally, the format list contains two values (or payload types): 102 and 98. From Table 12.5, we see these are dynamically assigned values (meaning look further in the payload for their definition). The order of these payload values is important, as they are listed in priority order for acceptance. More than two values may be provided, but two suffice for our example here.

Since these values are dynamically assigned, one must look to the attribute field (a=) to understand their definition. The first a= field identifies that the value 102 is a MPEG4 (MP4) codec at a clock rate of 90,000Hz. The second a= field shows the value 98 is mapped to an H.263 codec at a clock rate of 90,000Hz. Finally, a device can specify other attributes such as requiring a frame rate of 15 frames/sec.

21.3.2.4 Three-Way Conference Calling

The ability to conference multiple users together is a service that has taken many forms. A rich protocol such as SIP will allow many variations of a service that is only limited by the creativity of the designer. In this vignette, we looked at the most familiar form of conferencing—namely three-way conference calling. In the next two chapters, we will look at other forms of conferencing such as the conference bridge and a Push-to-X conference.

As indicated, SIP provides the flexibility to offer a service various ways to be implemented based upon the specific service definition, and three-way conference calling is an ideal example to demonstrate that flexibility. For instance, since SIP is a peer-to-peer protocol, if one had a very intelligent device with proper media resources, one could provide for the three-way bridging capability in the user's device as shown in Figure 21.20.

While the above call flow is a legitimate implementation, it has certain limitations that make it practical only under certain conditions. One key drawback is the additional complexities (translate additional costs) it adds to the end-user's device such as the requirement to bridge multiple channels (or media streams) together. In addition to potential additional service tariffs to implement this scenario, the additional device complexities may lead to a higher power consumption requirement, which is a limiting factor when it comes to mobiles devices. So, this flow would be valid within an office/PBX environment or a PC-based device, both of which would have its own power supply but most likely would not be practical for a mobile device.

Another limitation to a device-based solution (as is true with any device-based solution) is it is more difficult to take advantage of network-based information or to be blended with other applications. Thus, for our three-way conference calling example, we will look more closely at a network-based solution and introduce the concept of a conference server.

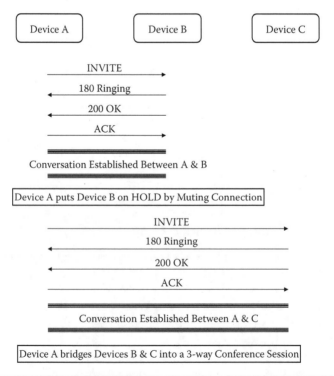

Figure 21.20   Device-based three-way calling.

Figure 21.21 and Figure 21.22 show the call flow for our three-way conference-calling example. We start examining the scenario from the point where Lauren has already established a call with Robert and now wants to add on Heath's parents. Our call flow follows the service description currently in use (i.e., pre-IMS, pre-VoIP) today for three-way calling, namely User A and User B are conversing and want to add User C. User A puts User B on hold, contacts User C to inform him of being added to a conference and finally bridges all three parties together. This service description makes use of a network-based conference bridge (typically colocated on the central office [CO] or mobile switching center [MSC]), which is analogous to our conference server[27] shown in the call flows as the AS/MRFC (as it would consist of these two functions).

In the first part of our call flow (steps 1–3), Lauren puts Robert on hold by sending him a second INVITE message with an indication to put this session on hold. When one needs to modify an existing session, a new INVITE request (also referred to as a re-INVITE) is sent within the same dialog used to establish that particular session. In this example, the media resources (RTP stream) are removed (or deleted) since a call on hold does not require these resources but the session is still maintained. A re-INVITE can modify other parts of a session such as changing

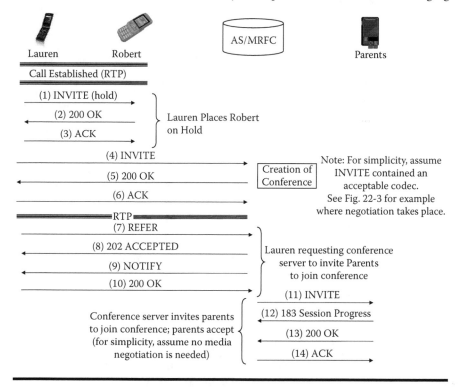

Figure 21.21 Three-way calling (part 1).

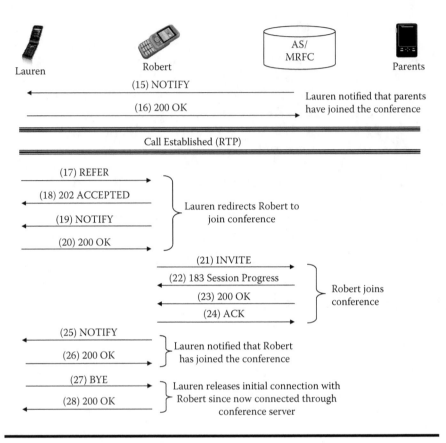

Figure 21.22 Three-way calling (part 2).

a port number or a URI, or add a new media resource as well as remove a media resource as shown in our example.

Once the first call leg is on hold, Lauren (also referred to as a conference participant) would request the creation of a conference session with a conference server (steps 4–6). She would then request the conference server to invite Heath's parents to join the call through the use of the REFER message (steps 7–10). Again, the REFER method is used by one SIP user agent to request the receiving SIP user agent to contact a third SIP user agent using the contact information supplied in the REFER message. In this case, the Refer-To URI header would contain a SIP URI indicating an INVITE message should be sent to request the third SIP user agent (i.e., Heath's parents) to join the conference (steps 11–16). What we have seen here is equivalent in the POTS[28] world of a user hitting a flash hook, entering the star (*) code for a three-way call followed by the number of the party to be added.

Once the call is established between Lauren and Heath's parents, they are able to converse briefly before adding Robert onto the call. When Lauren is ready to complete the three-way bridge, her UAC will send a REFER message to Robert's phone (UAC) redirecting (referring) it to the conference session (steps 17–24). After joining the conference, Robert will notify Lauren that he has joined the conference (steps 25–26). Now that all parties are connected through the conference bridge, Lauren's phone (UAC) will tear down the original session (currently on hold) established with Robert's phone since it is no longer needed (steps 27–28).

An astute reader may notice that two variations of connecting the participants to the conference server were shown: Robert sending an INVITE message to the conference server in response to a REFER message and Heath's parents responding to an INVITE message to join the conference. This is intended to show the flexibility of the SIP protocol in a service design. What is important here (in order to prevent interoperability issues) is that each device must be able to understand and implement the SIP protocol. Thus, it makes no matter whether a device is referred to the conference server (using the REFER message) or it receives an INVITE from a conference service to join a conference. Both devices implementing the standard SIP protocol will arrive at the same destination.

Endnotes

1. Rich voice would be a combination of plain voice with a data application or in combination with an enabler.
2. Typically, a network operator deploying one of these services a year is doing well. Mass market services are few, given the broadness of the target audience, not to mention the limited number of services the individual consumer can handle.
3. The mobile payment scenario is more of a silo service than a blended service. It is also typically invoked using near field communication (NFC) technology which is outside the scope of this book. It is mentioned here to call out the enhanced security that is inherent with the ISIM, plus it makes for a better story.
4. Implementation of the NAB using the XDMS structure is more synergistic with the IMS architecture than implementation as a stand-alone (non-XDMS) application just querying the Presence server. In the non-XDMS scenario, updates to the terminal could be done using SyncML or a proprietary format. However, as this book focuses on applications using the IMS, this will be the method detailed in the selected examples.
5. The original standard was developed as part of the Wireless Village Initiative, and hence the reference to it as Wireless Village. Wireless Village was later integrated into OMA and the work renamed to IMPS.
6. Phone directory for businesses.
7. Here we have assumed Bob has subscribed to the Personal Business Concierge service, a mythical local service provided by his town's business community, which businesses use to communicate with their customers about the status of their current dealings; in this case it would be an indicator (e.g., message or flashing icon) that Bob's pizza order is ready.
8. aGames is a hypothetical service that Bob's service provider is offering of a multiplayer gaming service.

9. Halo®3 is used only as an example of a well-known interactive game only. The authors have no knowledge of whether or not this game will be offered in any form for a mobile device by its manufacturer.

10. Different policies can be set up with different contact groups based upon need and willingness to pay. Service providers may choose to offer as a basic package only on/off status, whereas additional information such as access network type may only be provided with a higher package. Here we will assume that any higher cost is absorbed in the cost of Bob's aGames monthly subscription.

11. In reality, the user interface should present a differently named message (or action) for Bob to select such as a "My Buddies Status" button. Those in the know, know a SUBSCRIBE message is actually sent from the terminal device.

12. In the case of our example game, this would be a reasonable assumption for the players to have since as the game is described, IMS capabilities would be required.

13. For a service as described in this example, one would reasonably expect a time interval of a few seconds to a few minutes. Services with a low class of service level could be on the order of hours for their time interval.

14. We introduce a converged mobile device here. More in-depth examples will be provided in Chapter 24.

15. Most likely the message sent by the application server will follow TS.29.199-04 "Open Service Access (OSA); Parlay X Web Services; Part 4: Short Messaging." The details on Parlay X are outside the scope of this book.

16. 3GPP TS-24-141 "Presence Service Using the IP Multimedia (IM) Core Network (CN) Subsystem; Stage 3."

17. The aCalendar service is a hypothetical service which can be thought of as a Web-enabled calendar service that synchronizes with the mobile device. The user can set permissions to allow other people to set events or place reminders via a Web portal, which in turn would synchronize with the mobile device. In this scenario, it is a service provided to all students and faculty on the campus and works in conjunction with the mobile device.

18. The aCalendar Campus Alert is a hypothetical service that is a similar concept to the Personal Business Concierge application seen in Section 21.2 in that it provides a notification capability. However, in this scenario, it is more of a mass notice capability that is tied to a mass calendar application (e.g., look up for everyone placing the 8 a.m. history class on their calendar schedule).

19. With IMS messaging, the distinction between different types of messages—SMS, MMS, IM, or even voice mail—becomes blurred and is now more of a marketing distinction than a technology distinction. Thus, in this case, a voice mail is viewed as just another incoming message.

20. Video aShare is a hypothetical name for a one-way video streaming service that allows a real-time look at an event from one subscriber to another over a 3G IMS network.

21. The Short Message Peer-to-Peer (SMPP) Protocol is a protocol for network-based signaling between the SMSC and another SMSC or between an SMSC and a Short Message Entity (SME), which is defined as the source or recipient of a short message. In our example, the Parlay X Gateway would act as a proxy SME for the CCAS and connect to the legacy SMSC (either GSM or CDMA) using SMPP. As SMPP is not a defined southbound interface for a Parlay/Parlay X Gateway, this would be

a vendor-specific implementation that would reuse the Parlay framework for a non-Parlay-defined API. For further information on SMPP, the reader is referred to the SMPP Forum at www.smsforum.net.

22. The Video aShare service is analogous to the Video Share service offered by carriers today. We will use the more common name Video Share instead of the fictitious name Video aShare in this discussion.

23. Of course, there are other value propositions such as content adaptation, network interwork, and the ability to blend with other applications among other key benefits.

24. To further simplify the call flow, we have assumed only a single HSS in the network (thus there is no need to query the SLF for the subscriber's HSS ID) and the S-CSCF(A) is aware of I-CSCF(B)'s address (so there is no need to perform a DNS query to discover its address).

25. While a video streaming can theoretically work over a nonbroadband (i.e., 2G) connection, the user experience will not be very "customer satisfying." Therefore, most operators will choose to offer this type of service only over broadband connections.

26. As described in Chapter 12, Section 12.2.

27. As discussed in Chapter 2, we explored how the monolithic MSC was "broken down" into fundamental components to evolve into the IMS architecture. The conference bridge functionality provided by the MSC (or CO) is an example of an embedded service adapted into the IMS architecture. The service logic is moved into the AS, the control of the conference bridge ports is provided by the MRFC, and the bridge ports are provided by the MRFP.

28. POTS—Plain Old Telephone Service, a widely deployed pre-ISDN telephony system.

Further Reading

RFC 4353, "A Framework for Conferencing with the Session Initiation Protocol (SIP)," Rosenberg, J., February 2006.

RFC 3515, "The Session Initiation Protocol (SIP) Refer Method," Sparks, R., April 2003.

RFC 4579, "Session Initiation Protocol (SIP) Call Control—Conferencing for User Agents," Johnston, A. and Levin, O., August 2006.

OMA, "Push to Talk over Cellular," version 2.0, http://www.openmobilealliance.org.

OMA, "Presence SIMPLE," version 1.1, http://www.openmobilealliance.org.

OMA, "Converged Address Book," version 1.0, http://www.openmobilealliance.org.

OMA, "Instant Messaging and Presence Service," version 1.3, http://www.openmobilealliance.org

OMA, "Data Synchronization," version 2.0, http://www.openmobilealliance.org.

"GSMA Video Share Service Definition," March 27, 2007, http://www.gsmworld.com/documents/services/se41.pdf .

"GSMA Video Share Interoperability Specification," March 27, 2007, http://www.gsmworld.com/documents/ireg/ir74v1_1.pdf.

"Rich Communication Suite Initiative—Business Initiative White Paper," http://www.ericsson.com/technology/whitepapers/RCS_Initiative_White_Paper_RevA.pdf.

Chapter 22

Business Market Use Cases

The business (or enterprise) market space targets a different type of market segments from the consumer space. Whereas the typical consumer customer is more interested in convenience, social interaction, or simplifying one's lifestyle, the typical business customer is more focused on productivity. By definition, business-focused services tend to be more specialized than consumer services with a naturally smaller segment willing to purchase those services. Businesses are looking for applications that meet their specific business needs. Thus, IMS with its inherent tie-in to a rapid service delivery framework is a natural fit to create the niche services required by the business market segment in a rapid and economical manner.

For our "A Day in a Life" scenarios, we have created two vignettes targeting different aspects of the business market segment. Whereas in Chapter 21 we had all the services provided using a SIP AS (as would be expected in a pure IMS network), we introduce in this chapter services being offered from non-SIP AS but connecting into and being controlled by the IMS network. Our service blending examples here involve cross-technology applications from legacy wireless networks (CAMEL), Internet (Web services), and IMS (SIP) networks. By having the flexibility to pull applications from these different application technologies, the operator is able to create new service blends from already deployed networks (e.g., CAMEL) or to take advantage of an abundance of developers' toolkits (e.g., Web services) to meet their rapid or economical deployment requirements.

22.1 Office Professional

Scenario

Jim is an up-and-coming young attorney looking to faithfully serve his clients and impress his bosses. However, he has never liked working behind the conventional desk but prefers to be out of the office meeting clients or just working from his favorite local coffee house using his mobile office.

While meeting a client to do some investigation into some county courthouse records, he had to call back to his office paralegal staff to discuss a particular aspect of state law and have them do some research for him in order to know which documents to retrieve. Since Jim uses his mobile phone as an extension of his desk phone, it is capable of using his company's internal dialing plan, so he only has to dial the four-digit extension just like he would dial if he were at his office desk. He was glad that Mary picked up his call, as she was the most capable resource of the department. Since Mary was employed as a shared resource, her time needed to be charged back to a case as well as Jim's time had to be charged, ergo they had to log all their calls for billing purposes. Jim, being a little absent-minded along with being a little brash, couldn't be bothered with logging calls, either internal or external calls, so he was glad that his carrier provided a service that would log all his calls, do his record keeping for him, and provide him anytime access to those records.

After getting the information he needed from Mary, Jim was able to gather the documents he needed from the county records department and then make copies for his own record. He told his client he would have a report written up shortly and then headed off to his favorite coffee house to work. After his second espresso, Jim finished up his report and wanted to e-mail it quickly, as he knows his boss was anxious to get it. Being at the coffee house, he had to log into his corporate server remotely. Since his law firm dealt with a lot of sensitive cases, they had extra remote access security to prevent hackers from getting into their computer systems. Jim pulled up the remote access client and got the laptop microphone out. He spoke his name and today's date into the microphone and then hit the <Enter> key to initiate the connection. His corporate computer system has a voice recognition system that looks at all incoming requests. If a recognized voice and the appropriate date are not in the session request, the request is rerouted to some random URL to hide the system's existence from would-be hackers. After being recognized, Jim entered his standard login and password to access the corporate network. From there, he e-mailed his report to his boss and a senior partner before going back for a third cup of espresso, as he knew it involved a very important case.

Jim's report must have made a good impression, as before he finished his third espresso he had gotten an e-mail back from the senior partner asking him to set up a conference call in 30 minutes with all the partners. Seeing this request, his apprehension level was not helped by all the caffeine he had just drunk, but he knew how to set up a conference bridge, so he calmly sent out the meeting invitation with the bridge information and quickly moved to a quieter spot in the coffee house.

At the appointed time, Jim set up the conference bridge and conducted the meeting. Afterwards he got a surprise call from the senior partner congratulating him on a job well done and, after making a comment on how well Jim worked outside the office, he said there would be something extra on his desk if he ever came back to his desk. Hearing this, Jim rushed back thinking it might be a bonus paycheck. When he got to his desk, he just laughed when he saw his big bonus was a gift card to his favorite coffee house.

22.1.1 Scenario Applications

In our "Office Professional" vignette, we examined three different services and gave an example of two separate applications being blended together. We went beyond obvious, consumable services and looked at how IMS can be used to introduce an additional layer of security for the business supporting remote employee access. Finally, we will look at an alternative method to provide a related service we examined in the previous chapter. Thus, our services to be examined more closely are the following:

- Four-digit dialing with call logging
- Computer login with voice recognition security
- N-way Conference Bridge

22.1.2 Call Flows and Network Layout

We show our first use of the service broker in this first service example.[1] We will use the service broker to blend together two applications from two different technologies: CAMEL and SIP. Like many service broker products available today, we will assume the IM-SSF function is incorporated into the service broker. Our last two service examples will utilize standard SIP routing procedures, although our Remote Compute Login example will show a creative method of implementing the standard.

22.1.2.1 Four-Digit Dialing with Call Logging

Four-digit dialing is part of a suite of business applications offered as part of a virtual private network (VPN). An operator could have the VPN vendor integrate a call logging service into their VPN, but if the operator is unhappy with what the VPN vendor might charge or the quality of their product or possibly, as in this example, they are starting a transition to a new technology platform, they can look to a service blend to solve their dilemma. Figure 22.1 shows an example call flow for how the CAMEL SCP and SIP AS are blended together through the service broker to provide an end-user service. (Note that for our specific example, other SIP methods besides the INFO method could have been selected; INFO was selected for simplicity purposes only.)

The service broker here solves some additional problems in bringing these two applications together. First, in the transition to an IMS network, operators must still address a large legacy base (in our example here, a GSM/CAMEL-based network). Operators can start to enable these legacy customers to have access to IMS-based applications through the service broker. Second, a current limitation for a CAMEL network is the ability to trigger to multiple SCPs (or

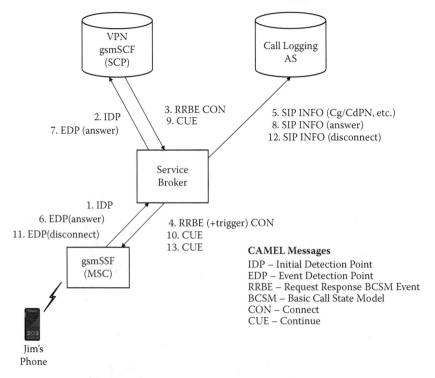

Figure 22.1   Four-digit dialing with call logging.

AS) from the same call trigger point (in our example, an originating call trigger). Here the service broker can provide a triggering capability to multiple applications servers in a manner similar to how the S-CSCF triggers to multiple application servers using the iFC. Finally, the service broker has the ability to provide additional service logic needed for the blended service. In this case, the CAMEL SCP (i.e., gsmSCF) would have no need to tell the GSM MSC (i.e., gsmSSF) to set a trigger notification for when the call terminates (as a four-digit dialing service is only concerned with providing a number translation service). However, in order to have the Call Logging AS receive a notice of when the call terminates (as Jim's law firm needs the full length of the call in order to bill its clients properly), we need the service broker to modify the response to the GSM MSC to tell it to set a call termination trigger. This last point of adding additional service logic controlled by the operator to a session is a key value that the service broker provides.

22.1.2.2 Remote Computer Login with Voice Recognition

Since SIP is a peer-to-peer protocol, it allows two endpoints to communicate directly and exchange its own control information. In this example, we make use of the SIP protocol support to carry a MIME payload with the INVITE message. A MIME payload is how an e-mail message carries an attachment and can be a text file, a graphic file, and/or a audio–video file. Some of the ideas behind including a payload with the INVITE message would be to include items such as a picture of the caller or possibly a company's logo. In our example here, the remote access client has the ability to attach an audio file when it attempts to set up a session with the host computer using the INVITE message (Figure 22.2).

The host computer upon receipt of the INVITE message can perform voice recognition analysis upon the payload, which could include some form of a verbal password (e.g., today's date) to add another level of security (instead of an always-repeated phrase that could be eavesdropped and recorded—for the paranoid among us). Hopefully, the client here is designed properly to delete the voice phrase after each use. Now upon successfully passing the voice recognition security level, our user can login normally.

We looked here at one specific example of how to use the payload capability in the INVITE message and mentioned a few other examples as well. One advantage is that IMS and SIP provide the ability to be creative in defining new services through the tools they provide.

22.1.2.3 N-Way Conference Bridge

An N-way conference call is a more general version of the three-way conference call that we looked at in Chapter 21. It is a commonplace service in the business world where N can be on the order of hundreds of conference call participants. In the

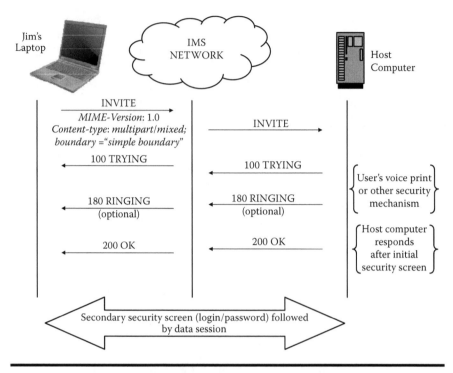

Figure 22.2 Computer login with voice recognition security.

IMS world for conference calling, there is always a central point where each conference call participant is connected to a central point of control (also referred to as a *focus*). This type of conference call is known as a tightly coupled conference. The focus is a SIP User Agent that is addressed through a conference URI to establish the signaling communication with the participants.

The conference call service has evolved from being a single media service (i.e., voice or audio) to a combination of media to support various meeting dynamics. Video conference calling is a common capability available in both the business world and with the home PC. Additional capabilities such as sharing a virtual whiteboard or file sharing are now available as well. We saw back in Chapter 12, Section 12.1 how the SDP description is used to designate both audio and video resources plus how one can set up a whiteboard even to the point of designating a landscape or portrait orientation.

For our example, we will assume that Jim only set up an audio conference call to simplify the call flows, understanding that similar procedures could be used to add the additional functionality. Figure 22.3 shows the call flow for setting up or creating a conference. The INVITE message sets the request URI to the conference AS (also known as the conference factory), so it may look like the following:

```
INVITE sip:conferencefactory@asmrfc@att.net SIP 2/0
```

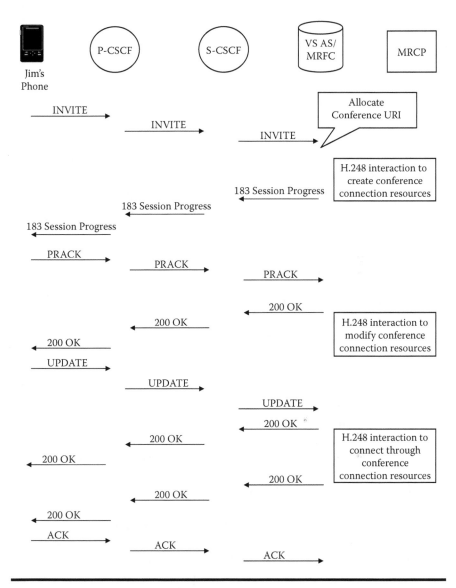

Figure 22.3 Conference call creation call flow.

Selecting the audio resources, the SDP portion of the INVITE message may show the media request as:

```
m = audio 2345 RTP/AVP 0
```

which is requesting a real-time voice (audio) circuit with a codec that supports PCMU (G.711). Again, if a video call is requested, then in addition to the above media attribute type, the SDP body would carry a media line that could look like this:

```
m = video 2552 RTP/AVP 34
```

which is requesting a real-time video path using a H.263 codec.

The other participants on the conference call would typically be given the call-in information ahead of time. Thus, they would be given the appropriate conference URI provided by Jim. The call flow for the participants would be similar to Jim's call flow for establishing the conference. In a way our N-way conference scenario is simpler than our 3-way calling example in the last chapter. Here, having the conference URI available to each participant minimizes the need to redirect call legs and allows the conference bridge server (or factory) to handle the bridging of the participants directly.

22.2 Field Force Worker

Scenario

It was finally a clear day after the past few days of thunderstorms and tornados. The Acme Insurance Claims Company was busy responding to all the new claims filings that were coming into the office. Marlene, the office manager, was busy trying to find available agents who could go to the actual damage site to file the insurance claims report. Fortunately, her company had the aDispatch[2] service that allowed her to see the location of the agent along with an indication of whether the agent was with a client or available to take a new claim filing.

While reviewing the different claims that needed to be investigated, Marlene saw that Hugh's status just turned to *available*. On her PC screen, his location just flashed green, while three other claim sites near Hugh's location started flashing. She quickly scanned the three claims and decided to send Hugh to the site that appeared to have limited damages, as he was still new to the company and lacked field experience.

When Hugh arrived at the claim site, he saw that a tree had been blown over and had fallen on the carport, causing heavy damage to the building's roof but only minor damage to car inside. After talking to the homeowners and getting some basic information from them, he went to inspect the damage. He took out his mobile phone and activated his aVideoLocker[3] service to record the scene for his company's records. He started on the outside of the building, shooting video footage while verbally describing the damage and worked his way to the inside of the carport. Hugh quickly finished up his inspection, did a final discussion with the homeowners, and then went to his car to finish up his report.

He called up Marlene and said he was e-mailing his report back to the office and was going to take a short lunch break. After hanging up, he thought he would call his old college buddy who had moved across the country, since he had the company phone with him and did not want to pay for the call himself. After dialing the number, he heard an announcement stating this phone was for official company business and that he needed to enter his override PIN for the call to complete. Reminded of the company policy of using the company phone for personal business, Hugh quickly hung up the phone and went in search of a restaurant.

22.2.1 Scenario Applications

In our "Field Force Worker" vignette, we looked at two new services and a third service that is a variation of a service examined in two other vignettes. We will continue to look at applications offered through non-SIP application servers and the blending of network enablers with an application. Our services to be reviewed here are the following:

- Device Resident Location Applications with Published Network Availability
- VideoShare storage
- Selective Outgoing Call Barring

The remote computer login application mentioned in this vignette could use the same procedures as described in the previous vignette and won't be repeated here.

22.2.2 Call Flows and Network Layout

So far it may appear that each application is deployed on its own individual application server (AS). We will introduce in our last service example the concept of a converged container, which is a SIP servlet container that can house multiple applications.

22.2.2.1 Device Resident Dispatch Service

In this example, we want to show how applications can reside on an end user's device and offer a service transparent to the operator's network, but at the same time make use of information that the network can provide. In our vignette, a dispatch service was described that also shows the field force worker's availability. Here, we will assume Hugh's mobile device has a built-in GPS receiver and publishes his availability to the presence server.

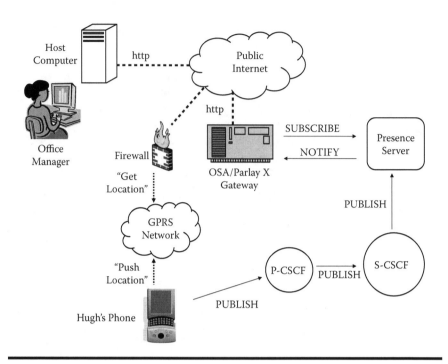

Figure 22.4   Dispatch service architecture.

Figure 22.4 shows the logical network architecture for our dispatch service. A business will typically want to host their own customer data and then query the network for supplemental information as opposed to hosting it in an operator's network.[6] Here, Marlene would enter the claims (including the address) into Acme's host computer so it has the location of the claims. Since our assumption here is that Hugh's mobile phone has a built-in GPS device, a separate data (http) connection can be established over the operator's data network (e.g., GPRS for a GSM carrier) to retrieve the device's location. Most likely, the device would have to periodically push its location to the host client although the host computer could make a request for the data. In a pre-IMS world, the mobile device typically would have to initiate a data session because there was no concept of an "always on PDP context." In a nutshell, a PDP (Packet Data Protocol) context is a data session established between the network and the mobile device. For numerous reasons beyond the scope of this book, it is only a temporary session.[4] With IMS, since all signaling is sent over a data connection, IMS devices have an "always-on" PDP context allowing a two-way flow of data anytime.

Acme's host computer would serve the role of a watcher and subscribe to Hugh's Presence (and Availability) status. As the request is coming from outside the operator's network, the request must go through the OSA/Parlay X Gateway as it must authenticate Acme's host computer. Depending upon the selected option, the presence server can provide the information upon a change in Hugh's phone status or when queried for the information by Acme's host computer.

For this scenario, we should state that this is not a practical implementation of a dispatch service with the current network technology. Instead, it is intended to show how an intelligent application can reside outside a carrier's network and make use of the information from within the carrier's network to provide a richer service experience. The creativity of an IMS network allows the reader to substitute their own applications in this scenario. Where this example fails to be practical is in two ways. First, while GPS receivers that can act independent of the operator's network were originally available in high-end smart phone models, the industry is quickly moving toward deployment of Assisted-GPS (A-GPS) in most 3G devices as a means to address emergency services requirements as well as location-based services. A-GPS has an advantage over a pure GPS device in that network data are sent to the mobile device to assist it in locating the satellites' positions versus a GPS device which must try to acquire the satellite's position on its own. Without these network-assisted data, the mobile device typically takes longer to find the radio signals from the various satellites (typically a minimum of three satellites) and can have other problems such as being in a downtown area (urban canyon effect) where high-rise buildings can obscure the radio signal. We will look at an A-GPS example in Chapter 23.

The other area where this example may have a flaw is that one might (correctly) assume if the device is intelligent enough to publish its own location, it should be intelligent enough to publish its own presence/availability status. If this were our scenario, then it would be a good example (although a boring example) of two highly intelligent endpoints (i.e., mobile device and host computer) handling all service logic and just using the operator's network for service data transport. For this case, the value that IMS provides is the "always on" PDP context so data can be freely exchanged between the two endpoints without the requirement to send wake-up messages.

It should be recognized that there are certain classes of applications that make more sense to be implemented transparent to the operator's network and to use the operator's network only for service data transport (also referred to as an "over the top" service). However, there are always trade-offs between implementations. "Over the top" services are limited in how they can work (e.g., only the host computer can access and make sure of the end-user's availability status) with other applications. This can lead to more expensive and complicated implementations if you want other applications to have access to this same information.

22.2.2.2 *Video Share to Recorded Session*

We looked at the Video Share service in the last chapter where it was a simple live video stream between peer devices with no application server between the end devices. As mentioned in that example, we assumed the current implementation of the Video Share service on a GSM network where the voice path is on a separate circuit switch connection from the video stream which is on its own data path connection. Here in our current example, our insurance agent needed to have his verbal notes recorded along with the video stream and not just the video stream itself. This is one of the drawbacks of the current GSM service, namely it is very difficult to the send the audio stream of a Video Share session to a storage locker because of the separate voice and data channels. This problem is eliminated if the IMS device and the IMS network support VoIP (Voice-over-IP). Using VoIP, both the audio and the video sessions can be combined into a single data path. In this case, both the voice and video sessions can be combined over a single data path. For our example, we will assume that Hugh has an IMS device that supports VoIP.

Figure 22.5 shows one logical network architecture view to support the audio–video session recording to a storage locker. Here, the INVITE message must make the request for needed network resources in the media-level description fields within its SDP parameters. One example SDP layout of the media fields is as follows:

```
m = audio 2345 RTP/AVP 0
m = video 7654 RTP/AVP 0
```

The media field (m) is repeated here to signify the fact that two different media types are being requested. The transport protocol <proto> RTP is required here due to the real-time nature of the session.

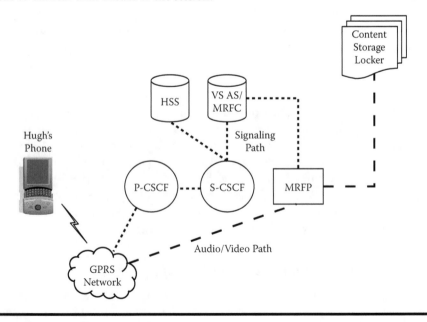

Figure 22.5   Video share storage.

In Figure 22.5, we have Hugh's mobile phone connected to the 3G RAN (radio access network), which connects to the carrier's GPRS data network. The SIP INVITE message is sent to the IMS core network (i.e., P-CSCF) for session initiation. In previous examples, we assumed the access network was implied and had the SIP signaling message go straight to the CSCF in order to simplify the figures. We show the transport network here in this figure to call out the point that the actual bearer path (i.e., audio–video stream) goes directly to the MRFP and not through the CSCFs.

The Video Share AS provides the role of the MRFC sending the control messages (using H.248) to the MRFP to establish the bearer path to the Storage Locker. Standards have been unclear to date on the separation of roles between the AS and the MRFC, thus one will see many implementations with the two elements combined.

22.2.2.3 Outgoing Call Barring

The network layout for the outgoing call barring service is shown in Figure 22.6. The INVITE message is routed to the AS where an analysis is provided on the called number (B number) found in the To: header to determine if it is on the permitted (or white) list.

```
From: "Hugh" <sip:+15123725555@ims.att.com;user=phone>;
tag=xy11t
To: +14255552345 <sip:+14255552345@ims.att.com>
```

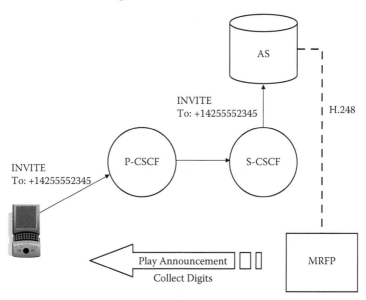

Figure 22.6   Call barring service.

Since in our example Hugh's college buddy is not on the permitted list (i.e., +1425-555-2345 is not a valid business number), his call attempt is intercepted for additional processing. The AS performs a dual role of the MRFC and sends a control message to the MRFP to play the intercept announcement and collect additional digits. Now one can get fancy with the service logic by now examining the calling party number (A number). If Hugh were president of the company as opposed to a new employee, then the service logic can be set up to look in the From: header field for <sip:+15123725555@ims.att.com> and any outgoing call would be allowed.

So far we have shown an application associated with an application server on a one-to-one basis. Although this is a legitimate deployment strategy, it may not always make an economical or an operationally wise decision. Especially as operators move into providing niche applications, it makes sense to share resources (both capital and support staff) among multiple applications. The Java standards community has defined a standardized approach for delivering SIP-based applications through the SIP servlet[5] API as specified in JSR 289. JSR 289 provides a known development environment that independent developers can use to offer their applications on a platform that can be shared with multiple applications. Figure 22.7 shows a conceptual layout of the SIP servlet application server as described in JSR 289. Here, we show the Call Logging and the Call Barring applications from our two Business Market vignettes along with some expansion capability for other applications.

In Figure 22.7, the container (or, to be more precise, SIP servlet container) provides the interface into the IMS (or SIP) network; receiving, responding, or initiating SIP messages with the IMS network. When it receives a SIP message, it determines which of its applications (SIP servlets) to invoke and the order in which to invoke them if more than one application is required. It will also manage the invocation of each application for the duration of the session.

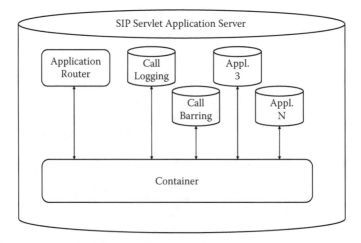

Figure 22.7   JSR 289 Application Server.

The Application Router (AR) receives the request from the container to service the incoming request from the IMS network. The AR is responsible for selecting the application requested by the container although it does not interact directly with the application. It may do so solely based on information received from the container or it may use external data such as a subscriber profile data base, time of day/day of week clock, service configuration data base, etc.

The Applications (such as our Call Logging or Call Baring examples) are Java-based application components managed by the container. These applications are typically independent of other applications and can be provided by different third-party developers with no knowledge of other applications executing on the same platform. Thus, not only can operators simplify their network topology by running multiple applications on a single platform, they now have a standardized mechanism to provide hosting services to third-party developers. Here, developers can deploy their Java-based application on the operator's application server and also have access to billing and subscriber provisioning functionality. This model is ideally suited to the host of small Internet developers who do not have the resources to host their own application servers.

Endnotes

1. Here we are using our broad definition of the SCIM as described in Section 6.4 to cover interworking with a SIP/IM network and a GSM CAMEL network.
2. The aDispatch service is a hypothetical mapping application that maps the location of the claim with the current location of the insurance agent along with the availability status of that agent.
3. The aVideoLocker service is a hypothetical storage locker application that saves an audio/video stream sent to it for future retrieval through a secure login process.
4. Mechanisms are available to "wake up" the mobile to establish a connection for downloads to be pushed to the mobile device. This usually requires a Push-Proxy Gateway to send a "wake up" message to the mobile device. The reader can explore this topic further by referring to the OMA Content Delivery specifications.
5. A SIP servlet can be viewed as a component of a Java-based application that interacts with the SIP servlet container.
6. Of course, there is the whole topic of cloud computing that could be mentioned, but that is for a future vignette.

Further Reading

3GPP TS23.078, "Customised Applications for Mobile Network Enhanced Logic (CAMEL) Phase 4, Stage 2."

3GPP TS 24.147, "Conferencing Using the IP Multimedia (IM) Core Network (CN) Subsystem; Stage 3."

OMA, "Dynamic Content Delivery," candidate version 1.0, http://www.openmobileal-liance.org.

JSR 289, "SIP Servlet Specification," v1.1.

Chapter 23

Converged Market Use Cases

Convergence is one of those terms that has many definitions to many different people, usually based on their particular perspective. To an engineer, it may mean the bringing together of different types of networks. To the salesman, it may mean bringing existing applications to new market segments. To the marketing organization, it may mean an opportunity to create an entirely new market segment. To the independent application developer, it may mean bringing the Web world to the wireless world. Many definitions exist, and they are all probably correct from a particular perspective. Here is one way that we inside of AT&T™ like to view convergence:

> An integration of traditional or new data/multi-media communications services in ways that add significant value for the end-user by making the communications easier or more transparent to use.

We will explore a combination of these definitions in this chapter as we look at our final two "A Day in a Life" vignettes. Our first story addresses the market segment "Families with Young Children." Here, we will look at the convergence of two different segments coming together to provide a specific solution by having a single application crossing multiple access technologies. We will also explore the role of the third-party developer working closely with the network operator to provide a real-life solution to a real-life need.

In our last story, we will look at an example of what has been called the "triple play" of telecommunications, a convergence of wireless, wireline, and IPTV applications into a seamless user experience.

23.1 Families with Young Children

Scenario

It was a warm spring day as Stephanie was pondering about working in the garden. Her two boys had been inside all morning playing video games, and she wanted them to go outside for some fresh air and exercise. With a typical preteen attitude they reluctantly turned off their video game player and went outside. "Mom," yelled Daniel as he closed the door, "I'm taking the IMS phone with me," thinking if he got really bored, he could access the aGames gaming server and try out a few new games. Nathan followed his big brother outside carrying a hammer and asking his brother if they could work some more on the tree house they were building in the woods behind their house. Seeing them working together on a project made Stephanie smile, as she told them to go ahead and work on the tree house.

About 30 minutes later while she was working in the garden, she heard her phone she had place in her garden tool bucket start ringing. It was Daniel. "Mom," said Daniel, "can you come get us? We don't know how to get home." It turns out that Daniel and Nathan had wandered off in the woods looking for material for their tree house and had come out of the woods via a wrong direction and were in a neighborhood they did not recognize. "Hold on a moment," Stephanie said exasperated. She quickly went to the menu screen on her phone and opened the child tracker service she had for her children's phone. After talking to Daniel for a few moments to make sure that he stayed calm, she had their location pinpointed. "How you have wandered off!" Stephanie said to Daniel. It turns out they were about a half-mile away from the house on the far side of an adjacent neighborhood. She told them to stay where they could see the road and she would be by in about 5 minutes to pick them up.

Stephanie got into her car and, following the map locator, she quickly found her two boys sitting by the street waiting for her. She noticed that Daniel had his arm around Nathan, trying to comfort him. Inquiring further she found that they had gone through some brushes and had gotten pretty scraped up. "Well, let's get you two explorers home and take care of those wounds," said Stephanie as she loaded her two boys in the car.

Upstairs in the bathroom as she was putting bandages on the scrapes, Nathan started itching and complaining about the scrapes on his legs. Looking closer, Stephanie saw a rash was developing and some hives starting to appear. Becoming alarmed now as this was more than just a few scrapes, she decided to call her doctor's office for advice. After she explained the situation to the nurse, the nurse replied, "Your family chart shows that you have the IMS Video Share service set up with our office. Can you show me a video picture of the affected area?" Quickly Stephanie added a video share link to the nurse's computer station by calling up the connection link that the doctor's office had provided in a previous visit. As Stephanie was streaming the video to the nurse's station, the nurse was directing her to move her camera phone around in different positions so that she could get different views of the affected areas on Nathan's legs. "Yes," said the nurse, "you were correct to assume it was poison ivy. You'll need to wash your hands immediately and get all their clothes in the laundry right away. I see your local pharmacy in our records. I'll call in a prescription right away. It should be ready by the time you get those boys of yours washed up and their clothes in the washing machine." Hearing all this, Stephanie was relieved that she didn't have to make a separate trip to the doctor's office. "OK, boys" said Stephanie, "Let's get you both washed up so we can go to the pharmacy. And if I get good cooperation from you, maybe we can pick up some ice cream along with Nathan's prescription." "See," said Daniel to Nathan, "I told you it would be fun exploring the woods."

23.1.1 Scenario Applications

In our "Families with Young Children" vignette, we saw two applications that combined a standard voice call with a simultaneous data service. We extended the service scope from the similar service discussed in Chapter 21 to add a PC screen as one of the endpoint device versus just mobile phone endpoints. Here, we introduce the "second screen" into offering IMS services. In this vignette, the services are as follows:

- Two-way voice call with a location tracking service
- Two-way voice call with a video share service to PC

As we saw in this use case scenario, there is a blurring of the lines between a consumer and enterprise application with the video share service. In fact, many converged type applications are expected to find use case applications that cross both market segments. In our example, the video stream went from Stephanie's mobile phone to the PC screen in the doctor's office, which (most likely in a real-life scenario) would be directly connected via DSL or to a DSL connection via a Wi-Fi hub.

23.1.2 Call Flows and Network Layout

In our network layout scenario, we will introduce the Location Services (LCS) architecture as defined in Third Generation Partnership Project (3GPP™). LCS is defined to work in both an IMS and non-IMS network. Here, we will look at its IMS attributes. We will also take a look at how IMS crosses access technologies boundaries to provide an end-to-end converged network experience.

23.1.2.1 Location Tracking Service with Voice Call

The child tracker service has been consistently listed in many market surveys as one of the highly desirable location-based services. For our example, we will assume the child tracker service is offered by a third-party service provider operating external to an operator's network. This third-party provider provides services to the operator's customers through the operator's Parlay X Gateway, which in turn provides the open interface for third-party application developers to offer their services (see Chapter 6.3). Also, in the breakdown of this service, we will introduce the Secure User Plane Location (SUPL) architecture, which is used to obtain the location information of the mobile device (also called the SUPL Enabled Terminal or SET in the SUPL specifications).

Figure 23.1 shows one implementation of the network architecture to support our child tracker service based on SUPL. We have already shown a voice call with a data session in previous vignettes, so we will not repeat the voice call here but instead we will focus on obtaining the location information component. For this

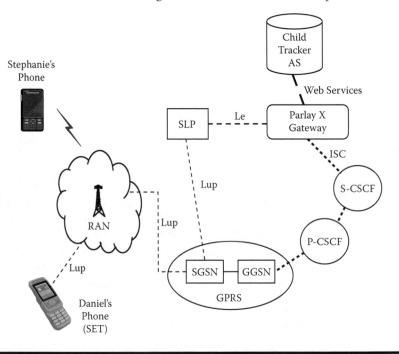

Figure 23.1   Child tracker network architecture.

example, we need to go beyond examining just the IMS network and look at the specific access network architecture. In Chapter 4.1, we introduced the GPRS network, which consists of the GGSN and the SGSN. SIP signaling goes through this access layer to get to the IMS network. Here, the client in Stephanie's phone will send an `INVITE` message toward the child tracker AS to establish a session. The `INVITE` is forwarded to the Parlay X Gateway where a protocol translation into a Web Services message is performed. The child tracker AS performs the usual service-level authentication and then sends a `[WS]GetLocation` back to the Parlay X Gateway to make the location request for Daniel's mobile phone. The entire set of Parlay X messages for obtaining the terminal's location can be found in 3GPP TS 29.199 part 9.

The OMA Location Working Group has identified two architectures for determining the location of the mobile device: Control Plane Location and the Secure User Plane Location. An in-depth look at these architectures is beyond the scope of this book; however, we will look a little deeper at the SUPL architecture as it will be the implementation used for location-based services. Although the Control Plane Location may also be used for location-based services, in practice it is mostly being implemented for use with emergency services only due to the high cost of implementing this architecture. By contrast, the SUPL architecture makes use of the existing GPRS network infrastructure to present a much lower cost implementation. It is expected that once VoIP becomes deployed over a mobile operator's network, then SUPL will assume the location aspect of emergency services as well for those end-users.

The SUPL architecture in its most basic form consists of three components (Figure 23.2). These three components are the terminal (or SET), the SUPL Location Platform (SLP), and the host application server. Both the SET and the AS host a location application that can request and consume the location information. The SUPL Agent is a service access point to request the location information through a network resource. The User Plane carries the exchange of information to determine the SET location along with any service management data. As was introduced in the previous chapter, most 3G devices are now coming with A-GPS

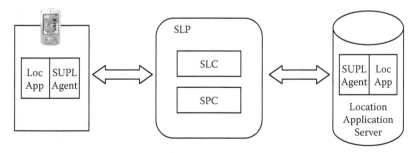

Figure 23.2   Basic SUPL architecture.

receivers in them to determine the device's location. A-GPS devices are provided with data from the network (in this case, this data is provided from the SLP) to aid in finding the positioning satellites.

The SLP consists of two functional components:

- SUPL Location Center (SLC)—Handles all service management functions such as setting up the session with the SET and establishing QoS parameters
- SUPL Positioning Center (SPC)—Addresses positioning calculations and obtaining and delivering the satellite positioning data for A-GPS receivers

OMA defines the Mobile Location Protocol (MLP) as the application interface between the SLP and the Location AS. However, as seen in Figure 23.1, this is not always true from an implementation point of view. If the Location AS is external to the operator's network, the entry point may or may not be MLP. Since operators will place a Parlay/Parlay X Gateway as an entry point to their network, they may expose a Parlay X Interface or they may build support for native MLP onto that gateway and expose a native MLP Interface to third-party developers. This decision will be left to each operator, although the momentum of the third-party development community (or ISV) will most likely slant the direction toward Parlay X, as that is a more well-known specification to the broader application development community.

The child tracker AS will take the lat/long coordinates it receives and translate them into a human-readable form such as a street address or point on a map. Depending on how the service is offered, a street map can be displayed on Stephanie's mobile screen, or an address with turn-by-turn directions can also be sent back for display.

23.1.2.2 Video Share Call between Mobile Phone and PC

This scenario differs from the scenario in Section 21.3.2.3 in that now one of the endpoints is a PC, versus both endpoints being mobile devices. In this earlier example, we showed a peer-to-peer case where the video stream flowed directly between similar devices on the same network. In our "Families with Young Children" vignette, the video stream was sent to a different network, and content adaptation must occur to modify the video stream to adapt to the display of the PC. In addition, for business or legal reasons, the doctor's office in our vignette may need to keep a permanent record of the video stream for future reference. The content adaptation and permanent recording will require an applications server with a media switch performing transcoding and a content storage repository (or locker). This differs from our earlier example in Chapter 21, which showed none of these elements because the service description did not require them.

Figure 23.3 shows one example implementation of the network architecture to support this service scenario.

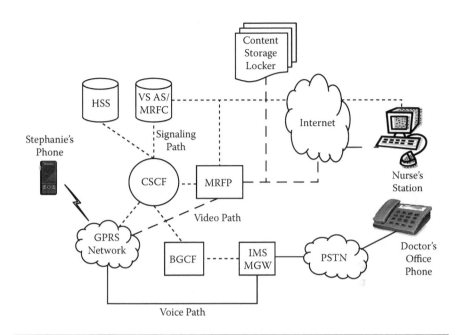

Figure 23.3   Mobile to PC video share call.

For our example, we can reasonably assume that both endpoints have an IMS Video Share client to facilitate communications. The knowledge of the doctor's PC URI can either be in the mobile client's address book or it can be located in the video share AS. A side note regarding the video share AS shown in our figure: To date, standards have not been clear on the clean separation of functionality between an AS and an MRFC. Because of this ambiguity, many implementations combine the two functions into a single element (in fact, many 3GPP specifications show them combined). The voice path would be established using normal, standard procedures. The video stream path would be established through the video share AS, which we will examine next.

When Stephanie initiates the video share session from her mobile phone, the INVITE would go first to the video share AS, which would play the role of a back-to-back user agent (B2BUA) toward the doctor's PC SIP client. In this role, it would negotiate with the PC for acceptable media connections (e.g., what codecs are supported) through the Offer/Answer model (see Chapter 12). Once the SDP negotiation successfully completes, the video share AS will direct the MRFP to apply the appropriate transcoding (e.g., H.263 AMR NB to MPEG4) and stream the video to the doctor's PC. It can also request the MRFP to send the stream into a content storage locker for retrieval by the doctor's office at a later date. The storage locker would be required to have the appropriate security mechanisms in place to ensure privacy rules are maintained.

23.2 Community Sports Fan Group

Scenario

Paul and his son Will are avid University of Texas Longhorn fans and spend their autumn weekends together watching their football game. Some weekends, Paul's alumni buddies and their sons come over to the house to watch the game on his large screen IPTV. This game, however, they are watching alone. Paul's friend Adam and his son Grant were lucky enough to get tickets to the game, while his other football friend David and his son Hayden are watching the game from their own house.

While they are watching the pregame show, Paul hears his cell phone ring and sees Adam's name displayed on the IPTV screen. For Adam, Paul will leave his comfortable lounge chair to retrieve his cell phone, as he also wants to show his friend all the good food he is missing. "Hey, don't pick up so quickly" is Adam's response to Paul answering his phone. "Your ring tone didn't get finishing playing our Alma Mata's fight song." "Don't worry," smirks Paul. "All the eyes of Texas will be on you today." Looking at his phone display, Paul sees that Adam is within 3G coverage and tells him that he is going to push him a video of the food spread he is missing. Adam agrees it looks real tasty but adds the stadium crowd is really full of energy and that he is going to push a video to show the crowd's excitement. Seeing the crowd definitely makes both Paul and Will wish that they were at the game.

"Hold on a moment" says Paul. "Let me see if David is watching the game." Using the set-top box (STB) remote, Paul pulls up his "Longhorn friends" community address book, which shows if his buddies are watching any Longhorn sporting event. He sees from David's presence status that he has his IPTV tuned to the football game. Just as he is about to set up a Push-to-Show session, he sees David's presence status turn to "OFF." "Hmm, that's strange," says Paul to Adam. "David switched off the game." They chat for a few more seconds when they see David's presence status change to "ON" again. "Hey, wait a second," Paul says. "David is back watching the game. I bet I know what happened. It was a commercial and David was probably channel surfing."

Adam calls up David and bridges him onto the call with Paul so he can join their video chat (Push-to-Show) session and pushes the video stream to both of them. David now sees the stadium crowd's excitement and wishes he had tickets, too. The three of them chat for a while until the game starts.

During the game, Paul receives a call on his home number from one of his wife's friends whose name appears on the caller ID screen on his IPTV. He knows he would only take a message if he answers the call, so he sends the call directly to voice mail. A little later, his wife Sarah calls. Seeing her caller ID come up on the IPTV screen, he knows he needs to answer but the game is at a crucial stage. Knowing better than to ignore her call, he answers, knowing that when he does answer the game is automatically paused and recorded for him so he won't miss a single moment of the game.

During the game, both the dads and the sons are messaging each other about great plays that they see. After a very exciting game in which the Longhorns pull out a last-minute win, they get a message from Adam to go to his blog page and see the pictures and videos he and Grant took. Paul and Will go to their PC to access Adam's blog and enjoy the game all over again, seeing the "in-person" pictures and responding back to Adam's blog.

23.2.1 Scenario Applications

In our "Community Sports Fan Group" vignette, we explored the extension of on-line communities of interest and buddy (friends) groups to incorporate IMS applications such as Push-to-Show and enabler functions such as Presence across a three-screen environment. For this vignette, we have included using IPTV as an IMS access medium to complete an example of all of the three-screen components. So, the new services we will be looking at in more detail for this vignette are as follows:

- Answer Tone with Caller ID Display on IPTV
- Push-to-Show Conference Bridge
- Network Address Book with Presence Status
- Video Storage Locker

23.2.2 Call Flows and Network Layout

Our last vignette explores how use cases can be implemented across the third screen (i.e., IPTV). We explore further the use of the service broker to do blending between IMS and non-IMS networks. Next, we examine how changing the media description in the SDP body allows for a much richer service experience as we upgrade our previously discussed Push-to-Talk to a Push-to-Show service. Finally, we show some extension services for the network address book.

23.2.2.1 Answer Tones with Caller ID Display on IPTV

This service looks at the blending of two individual services based on a single trigger action (i.e., incoming call). To make our example more interesting, we will assume a converged operator where the mobile operator portion is maintaining their legacy CAMEL-based Answer Tone infrastructure but is interworking with an IMS-based IPTV business unit. Here we will use the service broker element to act as a bridge to both technologies. Figure 23.4 shows an implementation for our blended service.

The service broker appears as an AS to the S-CSCF and receives the INVITE message indicating an incoming call for Paul's phone and containing the calling party information (i.e., Adam's information). The service broker has the service logic to know that for incoming calls to Paul's phone, it should invoke the answer tone service and deliver the calling line number information to Paul's IPTV. Upon receipt of the calling party's number, the IPTV AS can do a dip to the network address book (NAB) (or local copy) to translate the incoming number to the name as listed in Paul's address book. This example shows another benefit of the service broker flexibility as a programmable element. Because both of the blended services (answer tones and caller name delivery) are real-time sensitive, messages can be sent to both their application servers simultaneously to avoid any latency delays.

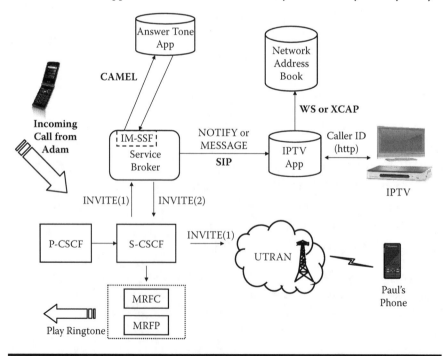

Figure 23.4 Blended CAMEL/SIP service.

This service could have been implemented through setting the iFC accordingly to invoke each AS. However, the iFC can only invoke multiple application servers in a sequential fashion that would introduce a delay on the second invoked service, which may cause an unwanted user experience since the S-CSCF must wait for a response from the first AS before it can invoke the second AS.

Many of today's implementations of a service broker have the IM-SSF function integrated into them to perform the SIP to CAMEL interworking for performance reasons. We will assume such is the case for this example. After receiving the INVITE message, the service broker performs a protocol translation to the corresponding CAMEL message—Initial Detection Point (IDP). Since the focus here is on IMS messaging, we will just note the appropriate CAMEL message and leave it up to the reader to investigate separately if more information about CAMEL messages is desired. The answer tone AS responds with an Establish Temporary Connection (ETC), which identifies the player for the designated ring tone. In our figure, note that the service broker must establish multiple sessions to address each of the call legs in our scenario (INVITE(1) to Paul's phone, INVITE(2) to establish the answer tone pathway). The service broker must have the service logic to associate the different INVITE messages. Thus it knows, when Paul answers his phone, to correlate the 200 OK(1) message with INVITE(2) in order to stop the ringback tone and to inform the S-CSCF that the call connection path should be completed.

For the IPTV Caller ID service, the service broker has a couple of options since it is just delivering an information payload without a requirement to set up a session. As shown a NOTIFY message or a MESSAGE message could be used to deliver the "A number" (Adam's number). An INFO message could also be used as an option as well.

There are several ways to obtain the calling name (aka "A number") information. Typically, a query is made to a database containing the billing name information (e.g., a Line Information Database [LIDB] as described in Telcordia GR-1158). An alternative method would be to query a NAB to obtain the known user associated with the particular phone. This is similar to today's mobile phone address book, which pulls up a personally designated name (e.g., "Mom") upon receiving an incoming call. Of course, if the NAB does not contain the entry, a fallback procedure to the LIDB would need to be in place. In cases of a family or the use of a company phone, the head of household or company name would appear if the LIDB approach is used. By invoking the NAB in the service flow first, a finer granularity of information (and more personal) is available to the called party over the LIDB approach. The connection between the IPTV AS and the NAB can take different forms based upon specific vendor implementations. Most likely, a Web Service would be used, although other methods such as XCAP could be used.

23.2.2.2 Push-to-Show Conference Bridge

We saw in our example how Paul and Adam were exchanging video pictures while still talking to each other. Here we want to show a variation of the Video Share and the Push-to-Talk service concept examined in earlier chapters, namely, Push-to-Show, to show one-way video being exchanged by both participants as opposed to only one participant sharing video. This is a natural evolution of the Push-to-Talk application currently available today. Here, instead of two (or more) users taking turns talking, they are taking turns sharing video streams using the floor control concept introduced with Push-to-Talk.

There can be many variations of this service concept depending upon the technology provided by the operator. Our example has Paul and Adam talking via a circuit-switched connection and pushing the video sessions via a separate data path. One can also have a combined Push-to-X service where the service allows users to combine both Push to Talk and Push to Video (Show) together. Another variation would have a full duplex VoIP session in combination with the Push-to-Show video. As we saw in our example in Chapter 22, Section 22.2, one of the key values here with VoIP is the ability to send the entire audio and video session together to a storage locker, which is not possible using a separate circuit switch voice path. The only alternative to get sound would be to capture the audio during the Push-to-Show session to obtain any ambient audio. However, in this use case, only the audio on the sending end would be captured. The final evolutionary end product might allow for full audio and video duplex transmission (technically, this last instance is not a Push-to-X variant, as these would be continuous streams). This use case has been dubbed "video telephony" and is what is available today, typically between PCs, but is expected to be available on mobile networks as higher broadband networks become available. Figure 23.5 shows the different service evolution concepts underlying the Push-to-Show service.

One Call Flow for a Push-to-Show implementation is shown in Figure 23.6 and Figure 23.7. We introduce here the Push-to-X Server, which is an enhanced version of the Push-to-Talk Server discussed in Chapter 18. The Push-to-X Server would typically handle all forms of Push-to-X services to allow for ease of integrating the different Push-to-X–type services (e.g., Talk, Video, Messaging, etc.) although it is not a hard requirement.

23.2.2.3 Network Address Book with Presence Status

A network address book (NAB) has the advantage of being shared across a user's multiple access devices without having to re-enter the contacts multiple times. In our example, Paul's address book is shared across his mobile phone and accessed by his IPTV service. When retrieving presence status information, a couple of approaches can be taken depending upon the intelligence of the end-user's device (e.g., cell phone or STB). If we view the STB functions as a Web browser (no

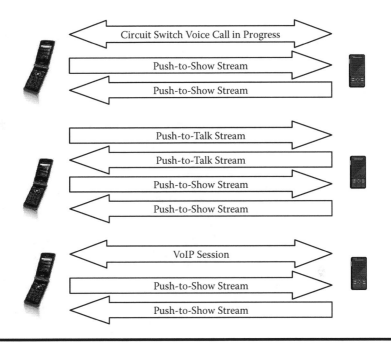

Figure 23.5   Push-to-Show variations.

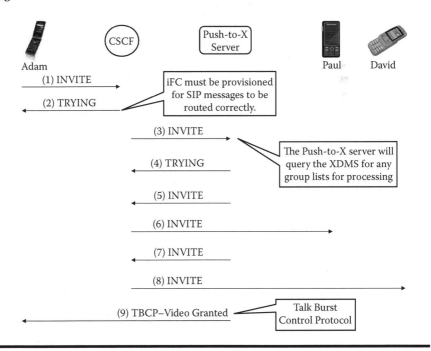

Figure 23.6   Push-to-Show call flow (Part 1).

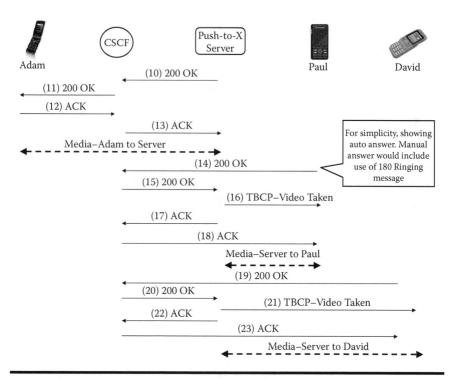

Figure 23.7 Push-to-Show call flow (Part 2).

storage of address book contacts locally but retrieved from the NAB on each query), then it will require the NAB to retrieve the presence information for it from the presence server (Figure 23.8).[1]

If the NAB synchronizes its information with the end device (keeps a local copy for performance reasons) such as Paul's mobile phone, then there is no need to go to the NAB to query the presence server, but instead the mobile phone (in this example) can send the query to the presence server directly (Figure 23.9).

For our vignette, since Paul pulled up his address book using his IPTV/STB, Figure 23.8 would be our most likely scenario here. In real-life deployments, the STB can typically act as a browser and go back "into the network" to retrieve their information such as the address book information or TV listings. For mobile phones, because of latencies in the radio network plus a desire to limit traffic over the air interface, implementations most likely will synchronize their data directly with the mobile phone. Different techniques can accomplish this data synchronization. Certainly a mobile phone can send a SUBSCRIBE message to retrieve updates via a NOTIFY message. XCAP can also be used to place the data appropriately into the contact files. Alternatively, OMA Device Synchronization (aka SyncML) can be used.

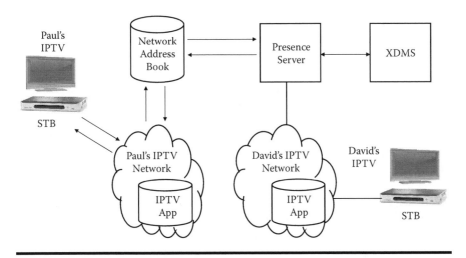

Figure 23.8   Alternative 1 for Presence-Enabled Address Book.

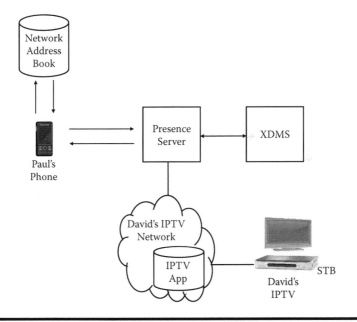

Figure 23.9   Alternative 2 for Presence-Enabled Address Book.

23.2.2.4 Video Storage Locker

When Adam uploaded his videos and pictures from his mobile phone to his personal blog page, it follows a similar scenario as our previous vignette. The video stream is forwarded to Adam's personal storage locker (which could be a Facebook™ or MySpace™ page) through the MRFP which can provide content adaptation and routing functionality. Others can access the blog over an HTTP stream via their mobile phone or PC.

Even though this particular service is limited in its IMS required functionality, what should be understood is how IMS technologies works harmoniously with other technologies (in this case, HTTP) to provide a unique end-user experience, which is the ultimate goal of deploying IMS.

Endnote

1. For our example here, we are assuming the NAB is separate from the XDMS. As discussed in Section 17.5, there is a study effort for OMA XDM v2.1 to use the XDMS as the address book repository. This example can be modified if necessary once the standard solidifies. Our example is currently reflective of what would be a real-life implementation.

Further Reading

3GPP TS 23.002, "Network Architecture."

3GPP TS 29.199 Part 9, "Open Service Access (OSA); Parlay X Web Services; Terminal Location."

Open Mobile Alliance, "PoC User Plane," approved version 1.0.2—05 September 2007, OMA-TS-PoC_UserPlane-v1.0.2.

List of Abbreviations

2G: Second Generation
3G: Third Generation
3GPP: Third Generation Partnership Project
3GPP2: Third Generation Partnership Project 2
AAA: Authentication, Authorization, and Accounting
ABMF: Account Balance Management Function
AF: Application Function
A-GPS: Assisted GPS
AIN: Advanced Intelligent Networks
AKA: Authentication and Key Agreement
AMPS: Advanced Mobile Phone System
AMR: Adaptive Multi-Rate
AMR NB: Adaptive Multi-Rate Narrow Band
ANSI: American National Standards Institute
AoR: Address of Record
API: Application programming interface
APN: Access point name
AR: Application router
ARIB: Association of Radio Industries and Business
ARPU: Average revenue per user
AS: Application server
ATIS: Alliance for Telecommunications Industry Solution
ATM: Asynchronous Transfer Mode
AuC: Authentication Center
AUID: Application Unique ID
AUTH: Authentication token
AUTN: Authentication network
AUTS: Synchronization token
AVP: Attribute Value Pair/Audio Video Profile
B2BUA: Back-to-back user agent
BGCF: Breakout Gateway Control Function

BoF: Bird of Feather
BSS: Billing support system
CAB: Converged Address Book
CAMEL: Customized Application for Mobile network Enhanced Logic
CAP: CAMEL Application Protocol
CAPEX: Capital expense
CCA: Credit-Control-Answer
CCAS: Campus Calendar Alert System
CCF: Charging Collection Function
CCR: Credit-Control-Request
CCSA: China Communications Standards Association
CDF: Charging Data Function
CDMA: Code Division Multiple Access
CDR: Charging data record; call detail record
CGF: Charging Gateway Function
CK: Cipher key
CN: Core network
CO: Central office
COI: Community of interest
CP: Client Provisioning
CPM: Converged IP Messaging
CPP: Calling Party Paid
CRLF: Carriage return/line feed
CS: Circuit switched
CSCF: Call Session Control Function
CSE: CAMEL Service Environment
CSN: Circuit switched network
CTF: Charging Trigger Function
DHCP: Dynamic Host Configuration Protocol
DM: Device management
DNS: Domain Name Service (or Domain Name System or Domain Name Server)
DS: Device Synchronization
DSL: Digital Subscriber Line
DTM: Dual Transfer Mode
ECF: Event Charging Function
EDGE: Enhanced Data rates for Global Evolution
ENUM: Electronic numbering
ESP: Encapsulating Security Payload
ETC: Establish Temporary Connection
ETR: Enabler Test Requirements
ETS: Enabler Test Specifications
ETSI: European Telecommunications Standards Institute
EV-DO: Evolution-Data Only

GERAN: GSM EDGE Radio Access Network
GGSN: Gateway GPRS Support Node
GPRS: General Packet Radio Service
GPS: Global Positioning Satellite
GRUU: Globally Routable User Agent URI
GRX: Global Roaming Exchange
GSM: Global System for Mobile Communications
GSMA: GSM Association
GUI: Graphical user interface
GUP: Generic User Profile
HLR: Home Location Register
HSDPA: High-Speed Downlink Packet Access
HSPA: High-Speed Packet Access
HSS: Home Subscriber Server
HTML: Hypertext Markup Language
HTTP: Hypertext Transfer Protocol
I/O: Input/output
IAB: Internet Architecture Board
IAD: Internet administrative director
IANA: Internet Assigned Numbers Authority
ICID: IMS Charging Identifier
I-CSCF: Interrogating-Call Session Control Function
iDEN: Integrated Digital Enhanced Network
IDP: Initial Detection Point
IESG: Internet Engineering Steering Group
IETF: Internet Engineering Task Force
iFC: Initial Filter Criteria
IK: Integrity Key
IKE: Internet Key Exchange
IM: Instant Messaging
IMPI: IP Multimedia Private Identity
IMPS: Instant Messaging and Presence Service
IMPU: IP Multimedia Public Identity
IMS: IP Multimedia Subsystem
IMS-MGW: IMS Media Gateway Function
IM-SSF: IP Multimedia-Service Switching Function
IN: Intelligent Network
IOI: Inter-Operator Identifier
IOP: Interoperability
IP: Internet Protocol
IPTV: Internet Protocol Television
IPv4: Internet Protocol version 4
IPv6: Internet Protocol version 6

IRTF: Internet Research Task Force
ISC: IMS Service Control
ISDN: Integrated Services Digital Network
ISIM: IMS Subscriber Identity Module
ISP: Internet Service Provider
ISUP: ISDN User Part
ISV: Independent Software Vendor
IT: Information technology
ITU: International Telecommunications Union
JSR: Java Specification Requests
K: Shared secret key
LCD: Liquid-crystal display
LCS: Location Services
LIDB: Line Information Database
LIA: Location-Info-Answer
LIR: Location-Info-Request
LTE: Long Term Evolution
MAA: Multimedia-Auth-Answer
MAP: Mobile Application Part
MAR: Multimedia-Auth-Request
MGCF: Media Gateway Control Function
MGCP: Media Gateway Control Protocol
MGW: Media Gateway
MIME: Multipurpose Internet Mail Extension
MLP: Mobile Location Protocol
MMD: Multimedia Domain
MMS: Multimedia Messaging Service
MMTel: Multimedia Telephony
MMUSIC: Multiparty Multimedia Session Control
MNO: Mobile network operator
MPLS: Multi-Protocol Label Switching
MRCF: Multimedia Resource Control Function
MRFC: Media Resource Function Controller
MRFP: Media Resource Function Processor
MSC: Mobile switching center
MSRP: Message Session Relay Protocol
NAB: Network address book
NAI: Network Access Identifiers
NE: Network Entity
NEP: Network equipment provider
NFC: Near Field Communication
NGN: Next Generation Networks
NNI: Network-to-Network Interface

NTP: Network Time Protocol
OAM&P: Operations, Administration, Maintenance & Provisioning
OCF: Online Charging Function
OCS: Online Charging System
OFCS: Offline Charging System
OMA: Open Mobile Alliance
OMTP: Open Mobile Terminal Platform
OP: Organizational Partners
OS: Operating system
OSA: Open Services Access
OSI: Open Systems Interconnection
OSS: Operations Support System; Operational Support System
OTA: Over The Air
PAG: Presence and Availability Group
PBX: Private Branch Exchange
PCEF: Policy and Charging Execution Function
PCG: Project coordination group
PCM: Pulse Code Modulation
PCRF: Policy and Charging Rules Function
P-CSCF: Proxy-Call Session Control Function
PDF: Policy Decision Function
PDG: Packet Data Gateway
PDP: Packet Data Protocol
PEAB: Presence Enabled Address Book
PIDF: Presence Information Data Format
PLMN: Public Land Mobile Network
PNA: Presence network agent
PoC: Push-to-Talk over Cellular
POTS: Plain Old Telephone Service
PPR: Push-Profile-Request
PRACK: Provisional Response ACKnowledgement
PRD: Permanent Reference Document
PS: Packet Switch; Presence Server
PSTN: Public Switched Telephone Network
PTT: Push-to-Talk
PTV: Push-to-Video
PUA: Presence user agent
PUID: Public User Identity
QoE: Quality of Experience
QoS: Quality of service
RADIUS: Remote Authentication for Dial In User Service
RAN: Radio Access Network
RAND: Random challenge

RCS: Rich Communication Suite
RES: Response
REST: REpresentational State Transfer
RF: Rating Function
RFC: Request for Comments
RLS: Resource list server
RPID: Rich Presence Information Data
RSS: Really Simple Syndication
RSVP: Resource Reservation Protocol
RTCP: RTP Control Protocol
RTP: Real-time Transport Protocol
RTSP: Real-time Streaming Protocol
SA: System Aspect
SAA: Security-Assignment-Answer; Server-Assignment-Answer
SAP: Session Announcement Protocol
SAR: Security-Assignment-Request; Server-Assignment-Request
SB: Service Broker
SCE: Service Creation Environment
SCS: Service Capability Server
SCIM: Service Capability Interaction Manager
SCP: Service control point
S-CSCF: Serving-Call Session Control Function
SCTP: Stream Control Transmission Protocol
SDF: Service Delivery Framework
SDO: Standards Development Organization
SDP: Service Delivery Platform/Session Description Protocol
SEG: Security Gateway
SET: SUPL Enabled Terminal
SGSN: Serving GPRS Support Node
SGW: Signaling Gateway
SHLR: Stand-alone HLR
SIMPLE: SIP Instant Messaging and Presence Leveraging Extensions
SIP: Session Initiation Protocol
SIPPING: Session Initiation Proposal Investigation
SLA: Service Level Agreement
SLEE: Service Logic Execution Environment
SLC: SUPL Location Center
SLF: Subscriber Locator Function
SLP: SUPL Location Platform
SME: Short Message Entity
SMPP: Short Message Peer-to-Peer Protocol
SMS: Short Message Service
SMSC: Short Message Service Center

SMTP: Simple Mail Transfer Protocol
SOA: Service-Oriented Architecture
SPC: SUPL Positioning Center
SPR: Subscriber Profile Repository
SQN: Sequence number
SS7: Signaling System Number 7
SSO: Single Sign-On
STB: Set-top box
SUPL: Secure User Plane Location
TBCP: Talk Burst Control Protocol
TDMA: Time Division Multiple Access
THIG: Topology Hiding Inter-network Gateway
TMF: Tele-Management Forum
TN: Telephone number
TP: Technical Plenary
TSG: Technical Specification Group
TTA: Telecommunications Technology Association
TTC: Telecommunications Technology Committee
TTL: Time-to-Live
TUI: Telephone User Interface
UA: User agent
UAA: User-Authorization-Answer
UAC: User Agent Client
UAR: User-Authorization-Request
UAS: User Agent Server
UDP: User Datagram Protocol
UE: User equipment
UI: User interface
UICC: Universal Integrated Circuit Card
UMS: User Mobility Server
UMTS: Universal Mobile Telecommunications System
UNI: User-to-Network Interface
URI: Uniform Resource Identifier
URL: Uniform resource locator
UTC: Universal Time Coordinates
UTRA: Universal Terrestrial Radio Access
VHE: Virtual Home Environment
VoIP: Voice over IP
VPN: Virtual private network
WAP: Wireless Application Protocol
WCDMA: Wideband CDMA
WG: Working group
WID: Work Item Description

WIN: Wireless Intelligent Network
winfo: Watcher information
WLAN: Wireless local area network
WSML: Web Services Mark-up Language
XCAP: XML Configuration Access Protocol
XCON: Centralized Conferencing
XDM: XML document management
XDMC: XDM client
XDMS: XML document management server
XML: eXtensible Markup Language
XMPP: Extensible Messaging and Presence Protocol
XRES: Expected response
XUI: XCAP User Identifier

Index

Note: page numbers *italicized* indicate figures or tables.